Advances in
Lipid Metabolism

Edited by Rodrigo Valenzuela Baez

Published in London, United Kingdom

IntechOpen

Supporting open minds since 2005

Advances in Lipid Metabolism
http://dx.doi.org/10.5772/intechopen.73796
Edited by Rodrigo Valenzuela Baez

Contributors
Fatiha Aid, Jin Yan Lim, Hiu Yee Kwan, Ljubica Tasic, Banny Silva Barbosa Correia, Raquel Torrinhas,
William Ohashi, Shahla Hashemi, Afnan Freije

Notice
Statements and opinions expressed in the chapters are these of the individual contributors and not
necessarily those of the editors or publisher. No responsibility is accepted for the accuracy of
information contained in the published chapters. The publisher assumes no responsibility for any
damage or injury to persons or property arising out of the use of any materials, instructions, methods
or ideas contained in the book.

First published in London, United Kingdom, 2020 by IntechOpen
IntechOpen is the global imprint of INTECHOPEN LIMITED, registered in England and Wales,
registration number: 11086078, 7th floor, 10 Lower Thames Street, London,
EC3R 6AF, United Kingdom
Printed in Croatia

British Library Cataloguing-in-Publication Data
A catalogue record for this book is available from the British Library

Additional hard and PDF copies can be obtained from orders@intechopen.com

Advances in Lipid Metabolism
Edited by Rodrigo Valenzuela Baez
p. cm.
Print ISBN 978-1-78984-458-0
Online ISBN 978-1-78984-459-7
eBook (PDF) ISBN 978-1-78985-137-3

We are IntechOpen,
the world's leading publisher of
Open Access books
Built by scientists, for scientists

4,800+
Open access books available

123,000+
International authors and editors

140M+
Downloads

151
Countries delivered to

Our authors are among the
Top 1%
most cited scientists

12.2%
Contributors from top 500 universities

CLARIVATE ANALYTICS
BOOK
CITATION
INDEX
INDEXED

WEB OF SCIENCE™

Selection of our books indexed in the Book Citation Index
in Web of Science™ Core Collection (BKCI)

Interested in publishing with us?
Contact book.department@intechopen.com

Numbers displayed above are based on latest data collected.
For more information visit www.intechopen.com

Meet the editor

Rodrigo Valenzuela Baez completed his pre-graduate studies (nutritionist) at the University of Chile (2003), his Masters in nutrition and food technology at the University of Chile (2007), and Doctoral degree in nutrition and foods at the University of Chile (2012). Currently, he is an Associate Professor in Nutrition Department at the Faculty of Medicine, University of Chile, Santiago, Chile. He does research in nutrition and food sciences. Dr. Valenzuela's teaching areas are food science and nutrition, and fats and oils in human nutrition. His current research areas are lipids in health and disease, metabolism and cytoprotection by long-chain polyunsaturated fatty acids from marine origin, bioconversion of n-3 and n-6 fatty acids from vegetable and marine oils, and physiological effects of n-3 fatty acid as a functional food. Dr. Valenzuela has more than 80 research publications in the lipid and metabolism area.

Contents

Preface   XI

Chapter 1   1
Roles of Lipids in Cancer
by Jin Yan Lim and Hiu Yee Kwan

Chapter 2   21
Analytical Tools for Lipid Assessment in Biological Assays
by Banny Silva Barbosa Correia, Raquel Susana Torrinhas,
William Yutaka Ohashi and Ljubica Tasic

Chapter 3   41
Plant Lipid Metabolism
by Fatiha AID

Chapter 4   57
Effect of Nanoparticles on Lipid Peroxidation in Plants
by Shahla Hashemi

Chapter 5   73
Fatty Acid Compositions in Fermented Fish Products
by Afnan Freije and Aysha Mohamed Alkaabi

Preface

This edited volume is a collection of reviewed and relevant research chapters concerning developments within the field of lipid metabolism. It includes scholarly contributions from experts in the field that cover such topics as roles of lipids in cancer, analytical tools for lipid assessment in biological assays, plant lipid metabolism, the effect of nanoparticles on lipid peroxidation in plants, and fatty acid compositions in fermented fish products.

The book includes chapters dealing with the topics: Roles of Lipids in Cancer, Analytical Tools for Lipid Assessment in Biological Assays, Plant Lipid Metabolism, Effect of Nanoparticles on Lipid Peroxidation in Plants, and Fatty Acid Compositions in Fermented Fish Products.

The target audience comprises scholars and specialists in the field.

IntechOpen

Chapter 1

Roles of Lipids in Cancer

Jin Yan Lim and Hiu Yee Kwan

Abstract

The term 'lipids' refers to a class of biological molecules primarily composed of hydrocarbons such as fatty acids, glycerolipids, sphingolipids and sterol lipids. Lipids take part in a variety of physiological functions and have specific roles depending on their chemical structure and localisation within or outside cells. For example, glycerolipids (e.g. triglycerides) are often used as energy stores, sterol lipids (e.g. cholesterol) and glycerophospholipids as structural components of cell membranes (e.g. the lipid bilayer), and sphingolipids as part of a signalling cascade. Since lipids are a source of energy and basic building block of all living cells, it is not surprising that development of cancer (i.e. uncontrolled proliferation of cells) is closely tied to the metabolism of lipids. This notion is supported by studies into the reprogrammed metabolic machinery in cancer cells, and also cell and animal model experiments showing that cancer growth and metastasis can be induced or inhibited by the exogenous addition of lipids. Here, we review how cancer cells can alter their lipid metabolism to meet their metabolic requirements, and the potential tumorigenic and tumour-suppressive mechanisms in which lipids are involved.

Keywords: lipids, cancer, metabolic reprogramming, signalling, autophagy, tumour development, cancer progression

1. Lipids in cancer

1.1 Lipid metabolism in tumours

Tumours can be simplistically described as masses of uncontrolled abnormal cellular growth. As they rapidly divide and proliferate, tumours require a steady source of energy and nutrients to accumulate biomass, and compete with healthy cells over a limited supply of essential cellular building blocks. Many cancers have adapted to their harsh environments by changing their metabolic profiles (the term 'reprogramming' is commonly used to describe this) to support growth and improve their chances of survival [1], among which the most well described is arguably their preference to perform glycolysis under aerobic conditions, an observation known as the Warburg effect [2]. In normal cells, glucose is hydrolysed via glycolysis, the tricarboxylic acid (TCA) cycle and oxidative phosphorylation to extract the maximum amount of energy in the form of adenosine triphosphate (ATP). This process utilises oxygen as the terminal electron acceptor during oxidative phosphorylation. In the absence of oxygen, glucose is still broken down to pyruvate via glycolysis, but is subsequently converted to lactate instead of being passed through the TCA cycle and oxidative phosphorylation. Metabolising glucose through glycolysis and fermentation into lactate results in smaller amounts of ATP compared to oxidative phosphorylation, however, tumour cells tend to prefer this path even in the presence of oxygen (i.e. the Warburg effect). A hypothesis to explain this preference suggests that instead of completely exhausting the carbon molecules in glucose through aerobic respiration (oxidative phosphorylation), highly proliferating cells need to conserve their carbon sources for the purpose of accumulating biomass [2, 3]. Vander Heiden and colleagues calculated the number of ATP and reduced nicotinamide adenine dinucleotide (NADH) molecules produced by glucose and compared these to the amount required for synthesis of macromolecules such as fatty acids, and concluded that proliferating tumours cannot utilise all their glucose stores for ATP production alone. The preference for glycolysis therefore could serve to increase availability of carbon-based precursors of biomolecules such as lipids, amino acids and nucleic acids that would otherwise be converted to carbon dioxide (CO_2) through respiration via the TCA cycle and oxidative phosphorylation.

By reducing the loss of carbon through respiration, tumour cells can utilise this saved pool for synthesising basic cellular building blocks necessary for sustaining their proliferation. One such example is the synthesis of fatty acids and other lipid molecules derived from the modification of fatty acids. Fatty acids and their derivatives have indispensable roles in cell biology; a few key functions include formation of the basic structure of the cell membrane, as an energy storage pool and as mediators in cellular signalling cascades. Lipids are typically obtained from dietary sources or synthesised in living cells beginning from the precursor molecule acetyl-coA. In most eukaryotic cells, pyruvate is produced from the breakdown of glucose via glycolysis. It is then funnelled into the mitochondria in which the enzyme pyruvate dehydrogenase converts pyruvate to acetyl-coA. Acetyl-coA is subsequently converted into citrate by citrate synthase (first step in the TCA cycle), a step necessary to transport acetyl-coA in the form of citrate from the mitochondria into the cytosol which is the site of fatty acid synthesis. Citrate is transported out of the mitochondria and converted back into acetyl-coA by ATP citrate lyase (ACLY) in the cytosol. Next, acetyl-coA is carboxylated by acetyl-coA carboxylase (ACC) to form malonyl-coA, and both precursors are then attached to an acyl carrier protein and repeatedly elongated with units of carbons from additional malonyl-coA molecules. This elongation is performed by fatty acid synthase (FASN) to produce a 16-carbon molecule termed palmitic acid. Palmitic acid can be further desaturated and/or elongated to produce unsaturated fatty acid derivatives which serve as building blocks for the synthesis of other lipids such as phosphoglycerides, phosphoinositides, eicosanoids and sphingolipids (summarised in **Figure 1**, reviewed in [4]. Separately, acetyl-coA is also used for the synthesis of cholesterol through the mevalonate pathway. This process involves first converting acetyl-coA into lanosterol (via intermediates including 3-hydroxy-3-methylglutaryl coA, mevalonate, isopentenyl pyrophosphate, farnesyl pyrophosphate and squalene), which is then transformed into cholesterol through a multi-step enzymatic process.

Studies have indicated that the biosynthesis of basic cellular building blocks including proteins, fatty acids and nucleic acids is modified and/or upregulated in [5, 6], indicating that the metabolism in highly proliferating cancer cells is likely altered to support their abnormal growth. Lipids and fatty acids in particular are required for the biosynthesis and modification of the lipid bilayer membrane in newly formed cells [7], and also for other roles related to cell signalling and tumour survival. Consistent with the fatty acid biosynthesis pathway, tumours primarily obtain carbon acyl fatty acid precursors from glucose [8, 9]. To increase the production of fatty acids and other lipids, tumour cells hijack the fatty acid biosynthesis pathway to their advantage. Component enzymes in the pathway (ACLY, ACC and FASN) are commonly upregulated in tumours [10–13], and inhibition or silencing

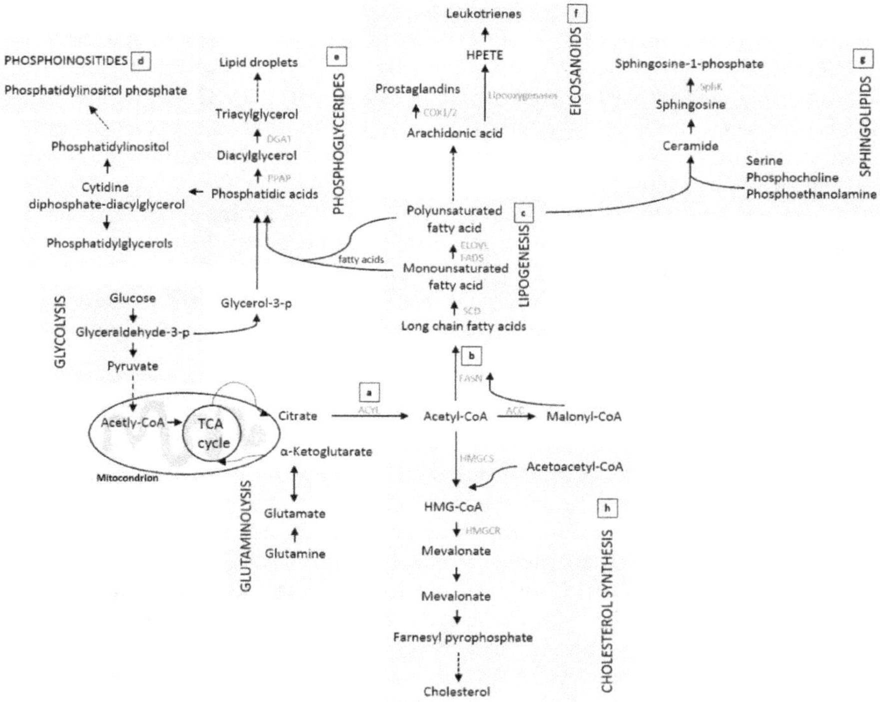

Figure 1.

Lipid biosynthesis. Schematic representation of the pathways involved in the synthesis of fatty acids, cholesterols, phosphoglycerides, eicosanoids, and sphingolipids. Enzymes involved in catalysing the process are labelled in red. (a) Citrate derived from the tricarboxylic acid (TCA) cycle is first converted to acetyl-CoA by ATP citrate lyase (ACYL). (b) For fatty acid synthesis, acetyl-CoA carboxylase (ACC) adds a carboxyl group to convert acetyl-CoA to malonyl-CoA. Repeated condensation of acetyl-CoA and malonyl-CoA catalysed by fatty acid synthase (FASN) results in a 16-carbon fatty acid chain. After which, the 16-carbon fatty acid chain is cleaved by thioesterase to generate long chain fatty acids such as palmitic acid, stearic acid, and oleic acid. The addition of a double bond by stearoyl-CoA desaturase (SCD) yields monounsaturated fatty acids. (c) Subsequent elongation and desaturation, catalysed by enzymes fatty acid elongase (ELOVL) and fatty acid desaturase (FADS) produces a pool of fatty acids with different saturation levels. Essential fatty acids can also be obtained from dietary intake. (d–g) Subsequent modification generates different types of lipids. (d and e) In glycerolipid biosynthesis, saturated and unsaturated fatty acids combine with glycerol-3-phosphate, a reaction highly dependent on glycerol-3-phosphate acyltransferase (GPAT) to generate (d) phosphoinositides and (e) phosphoglycerides. (f) Eicosanoids are signalling molecules made by oxidation of polyunsaturated fatty acids such as arachidonic acid. Downstream, multiple families of eicosanoids such as prostaglandins and leukotrienes can be generated. (g) Sphingolipids contains acyl chains and polar head groups derived from serine, phosphocholine, and phosphoethanolamine. Ceramide, sphingomyelin, and sphingosine are common intermediates of the sphingolipid metabolic pathway (h) cholesterol synthesis is regulated by a series of conversion and addition of acyl groups by enzymes 3-hydroxy-3-methylglutarate-CoA synthase (HMGCS) and 3-hydroxy-3-methylglutaryl-CoA reductase (HMGCR). Subsequent modifications yield farnesyl-pyrophosphate, an important intermediate for protein prenylation. Cholesterol also forms the structural backbone of hormone synthesis in the cell. Abbreviations used in the figure: coenzyme A (-CoA), prostaglandin-endoperoxide synthase (COX1/2), diacylglycerol O-acyltransferase (DGAT), 3-hydroxyl-3-methyl-glutaryl-coenzyme A reductase (HMG-CoA), arachidonic acid 5-hydroperoxide (HPETE), and phosphatidic acid phosphatase (PPAP).

of these enzymes has been demonstrated to restrict growth of cancerous cells [14–16]. The upregulation of these fatty acid synthesis-related enzymes is achieved through signalling by the mammalian target of rapamycin (mTOR) complex 1 and transcription factors called sterol regulatory element-binding proteins (SREBPs). SREBPs exert transcriptional control over various fatty acid, cholesterol, triglycerides and phospholipid synthesis and uptake genes [17] and are regulated by mTOR complex 1, a nutrient and growth factor responsive kinase [18]. Previous studies conducted in various cancers have implicated deregulation of mTOR signalling in mediating proliferation of cancer cells (reviewed in [19, 20]). More specifically, mTOR and SREBPs have been shown to increase lipid biosynthesis through Akt signalling thereby promoting proliferation in cancer cells [21]. The signalling by mTOR complex 1 also leads to upregulated fatty acid biosynthesis in cancer cells either by activating SREBPs via S6 kinase [22] or phosphorylating (downregulating) the SREBP inhibitor Lipin 1 [18]. In addition to lipids, mTOR complex 1 signalling is also implicated in promoting biosynthesis of proteins and nucleotides [23–25]. Taken together, these findings indicate that the deregulation of mTOR complex 1 plays a central metabolic role in promoting growth and proliferation of cancer cells by allowing them to 'reprogram' their metabolism. Indeed, there are studies into the potential use of mTOR inhibitors as cancer therapy drugs given its importance in the development of cancer.

1.2 Lipids as promoters of cancer

Early experiments have established that the lipid composition of tumour tissues is distinct from normal healthy cells [26–29]. Their lipid composition differs depending on the type of tumour tissue and possibly also correlates with tumour stage and malignancy characteristics, as recently demonstrated in a comparison of membrane lipid composition between six human breast cancer cell lines and healthy mammary epithelia [30]. These and other similar studies led to the notion that lipids could play an active role in cancers in addition to their basic function of maintaining structural integrity of the lipid bilayer membrane. One such example is a class of lipids termed sphingolipids. Sphingolipids are lipid molecules that contain an amino alcohol group in their backbones, and depending on additional substitutions with fatty acid residues or phosphocholine, form sphingolipid derivatives such as ceramides and sphingomyelins. The basic role of sphingolipids is to augment fluidity and barrier function of the lipid bilayer cell membrane in which they normally reside in the outer leaflet. Sphingolipids, in particular sphingosine-1-phosphate (S1P), have been demonstrated to promote cell survival during tumorigenesis as inhibition of either upstream fatty acid or specifically sphingolipid synthesis restricts tumour growth [31]. Sphingosine can be synthesised from condensation of palmitic acid with the amino acid serine, or from the cleaving of fatty acid residues from ceramides by ceramidase. The resulting sphingosine is phosphorylation by sphingosine kinase, producing S1P. S1P signalling interacts with histone deacetylase 1, 2 (HDAC1 and HDAC2) and telomerase to control many key cellular process involving cellular growth, proliferation, migration and invasion (reviewed by [32, 33]; see section below on lipids as signalling mediators in cancer), thus its metabolism and related enzymes are an area of considerable research interest.

A second aspect to the role of lipids in promoting cancer is the influence of exogenous sources of lipids in facilitating tumorigenesis and metastasis. Numerous studies have experimented with high lipid content diets using mouse models and reported increases in tumour growth and/or metastasis, implicating high fat ketogenic diets [34–36] or specific lipids such as cholesterol [37] or palmitic acid [38] in promoting cancer. There is a variety of mechanisms by which high concentrations of dietary lipids can exert a tumorigenic effect. According to Liśkiewicz and colleagues, their high fat ketogenic diet administered *ad libitum* to mice led to activation of ERK1/2 which controls cell proliferation, differentiation and survival [39], as well as elevated mTOR signalling in renal tumours [34]. In a different study, high fat diets caused acetoacetate levels in the serum of recipient mice to increase, subsequently leading to enhanced tumour growth of xenograft human melanoma cells with a V600E mutation in the BRAF gene [35]. Another mechanism by which high fat diets could enhance tumour metastasis is through the Ras-Raf-MEK-ERK mitogen-activated protein kinase (MAPK) pathway which was recently shown to activate SREBPs and therefore lipogenesis in metastatic human prostate cancer [36]. More examples of specific lipid groups linked to cancer include cholesterol and palmitic acid as mentioned above. The introduction of excess cholesterol either through dietary sources or by genetically increasing cellular cholesterol biosynthesis stimulated growth of intestinal crypt cells, leading to a more than 100-fold increase in the rate of tumour formation in the gastrointestinal tracts of live mice [37]. Similarly, exogenous addition of palmitic acid was shown to increase the invasiveness of human pancreatic cancer cells via a toll-like receptor 4 (TLR4)-mediated pathway [40], promote growth of melanoma cells through Akt signalling [41], and also increase the metastatic potential of human oral carcinoma through membrane-bound fatty acid receptors termed CD36 [38]. These studies collectively suggest that excess dietary lipids are detrimental to health and could exacerbate cancers in addition to obesity; however, whether these findings translate into appreciable risks of cancers in humans remains an open question.

1.3 Lipids as suppressor of cancer

On the other hand, not all classes of lipids appear to stimulate cancer growth and metastasis. There is evidence supporting an inhibitory role of polyunsaturated fatty acids (PUFAs) in cancer development [42–44]; reviewed in [45], although conflicting experimental results do exist [46]; reviewed in [47]. Dietary PUFAs commonly consumed by humans encompass two major groups—the n-3 and n-6 families of PUFAs. These PUFAs are categorised by the position of their first double bond from the methyl end of the fatty acid molecule (n-3 signifying double bond between third and fourth carbon atom, n-6 between sixth and seventh carbon atom). Some common n-3 PUFAs include alpha-linolenic acid (ALA), eicosapentaenoic acid (EPA) and docosahexaenoic acid (DHA), and common n-6 ones include linoleic acid (LA) and arachidonic acid (AA). The cancer promoting or inhibitory effects of PUFAs is hypothesised to depend on the relative amounts of n-6 and n-3 administered [48]. Current trends suggest that n-3 PUFAs are beneficial towards reducing cancer, whereas n-6 PUFAs tend to increase risks. An epidemiological survey tracking more than 72,000 female participants and their diets over an average duration of 8 years indicated that individuals consuming higher amounts of n-6 PUFAs relative to n-3 faced increased risks of developing breast cancer [49]. These trends in a large cohort were consistent with previous assessments of the beneficial properties of the n-3 PUFAs EPA [50–52] and DHA [53, 54] in fighting various cancers. The beneficial properties of ALA (also n-3), however, is less established compared to EPA and DHA. Consumption of ALA in mouse models of prostate cancer were shown to reduce cancer growth [46], although another study conducted on human prostate tissue presented evidence that ALA in the prostate was associated with aggressive prostate cancer [47]. The n-6 PUFA LA is commonly studied in the context of breast cancers, although its role is still currently unclear as studies of LA and risk of breast cancer have returned inconsistent results [55, 56]. The other n-6 PUFA, AA, is often studied in the context of prostate cancers and have been shown to increase prostate cancer growth [57, 58], although a meta-analysis of AA and the risk of various cancers including prostate only show weak associations [59]. The exact role of PUFAs in cancers most likely depends on many other factors including cancer cell type, stage and host metabolism of these PUFAs, all of which should be explored in more detail to exploit PUFAs in anticancer therapy.

2. Lipids as signalling mediators in cancer

Many cellular signalling hormones and growth factors have structural components comprising of lipids. Examples of such hormones and factors include prostaglandins, lysophosphatidic acid, and steroid hormones to name a few. Lysophosphatidic acid is a phospholipid derivative that binds G protein coupled receptors (GPCRs) to activate cell proliferation, survival, and migration. As such, tumorigenesis and cancer expansion is commonly attributed to dysregulated lysophosphatidic acid expression and signalling [60]. In addition, autotaxin, a secreted enzyme involved in production of lysophosphatidic acid is associated with hyper proliferation [61] and tumour invasiveness [62]. Overexpression of autotaxin and lysophosphatidic acid receptors was reported in several cancers including glioblastoma [63], prostate [64], and breast cancer [65], all of which overexpression contributed to increased cell motility and invasive potential. Notably, production of either autotaxin or lysophosphatidic acid receptors was sufficient to induce development of high frequency invasive breast tumours [60]. In human liver cancer cells, lysophosphatidic acid has also been shown to bind lysophosphatidic receptor 1 to activate MMP-9 signalling and promote cancer cell invasion [66].

Bioactive sphingolipids form an important class of lipids consisting of sphingosines, ceramides, and other complex sphingolipids such as sphingomyelins and glycosphingolipids. They bind specific protein targets to elicit signalling responses in important cellular events such as growth regulation, cell adhesion, migration, apoptosis, and inflammation [67]. Sphingolipids and its derivatives have been implicated in the regulation of signalling cascades in multiple aspects of cancer pathogenesis and therapy, in either tumour suppression or survival of various cancers [33, 67]. For instance, ceramides are commonly known to suppress tumour growth by mediating cancer cell death via apoptosis, necroptosis or mitophagy [68]. They are synthesised in response to cellular stresses that produce apoptotic signals such as chemotherapy or ultraviolet (UV) radiation [69]. Various modes by which ceramide regulates apoptosis have been proposed. One such example is in radiation-induced apoptosis, during which ceramide channels activate mitochondrial apoptosis through mitochondrial outer membrane permeabilization [70]. On the other hand, S1P is considered to be a pro-survival lipid as it is able to initiate cancer cell proliferation, malignant transformation, prevent apoptosis, and promote resistance to anti-cancer therapies [68, 71, 72]. SIP mediates host-cancer cell communication by engaging G protein-coupled S1P receptor-dependent or -independent signalling to promote tumour migration, survival, and evasion of host immune responses [73].

Prostaglandins are a subclass of eicosanoids. They are synthesised by the oxidation of 20-carbon essential fatty acids catalysed by phospholipases and cyclooxygenase (COX) enzymes. Prostaglandin E2 (PGE(2)) is the most widely studied and has been proposed to directly modulate tumorigenesis in several cancers (reviewed in [74]). For instance, administration of exogenous PGE(2) to F344 rat models resulted in higher incidences and multiplicity of intestinal adenomas [75]. Enhanced colon carcinogenesis was proposed to occur through the activation of PGE(2) signalling, by binding of E-prostanoid (EP) membrane receptors 1–4 [75]. A separate *in vitro* study showed that PGE(2) treatment upregulated epithelial cell proliferation and COX-2 expression in intestinal adenomas, proposed to act via the Ras-mitogen-activated protein kinase signalling pathway [76]. Other than PGE(2), uncontrolled expression of EP has also been reported and as a result affects the outcome of various cancers [77, 78]. For example, Jin and colleagues [79] demonstrated that activation of PGE(2) with EP1 receptor agonist ONO-DI-004, but not antagonist ONO-8711, improved cell viability and migration of liver cancer cells. In Lewis lung carcinoma cells, EP3 was shown to trigger production of MMP-9 and

VEGF, both of which are central regulators of angiogenesis and subsequent metastasis [80], further indicating the role of prostaglandin signalling in cancer progression. Taken together, the modification of signalling pathways by cancer cells affects abundance and activation of signalling lipids which, as a result promotes prooncogenic pathways that could lead to resistance against anti-cancer treatments.

3. Lipid-based post-translational modification of proteins in cancer

The understanding of the role of lipids in the modulation of cellular processes in cancer cells (with comparison to normal cells) is important to help identify potential cancer markers. Since post-translational modification of proteins is an important component in many key signalling components during oncogenic progression, they are a suitable candidate for cancer studies. Ongoing research has highlighted the importance of various post-translational modifications that contribute to oncogenesis, namely phosphorylation, glycosylation, ubiquitination, prenylation, methylation and acetylation [81]. A common involvement of lipids in post-translational modification is known as prenylation. Prenylation is a process in which a hydrocarbon-based hydrophobic group (such as farnesyl [a 15-carbon isoprenoid] or geranylgeranyl) is covalently attached to a protein post-translation, which as a consequence changes cellular localization, protein-protein interaction, and function of the modified protein [82]. Prenylation is crucial for membrane association and activation of GTPases such as Ras, Rho, cdc42, and GPCRs, all of which are important regulators of cancer [83, 84]. For instance, stimulation of Ras proteins is known to promote oncogenesis by regulating gene expression, cell cycle progression, survival and migration [85]. Inactivation of the retinoblastoma protein (a tumour suppressor protein) induced unregulated expression of farnesyl diphosphate synthase and prenyltransferases, subsequently increasing prenylation/activation of N-ras in retinoblastoma tumour and promoted senescence [86]. Furthermore, prenylation is also known to involve farnesyl-pyrophosphate, an intermediate for cholesterol synthesis. Given the importance of lipid-based post-translational modification of proteins, many anti-cancer therapies currently target proteins and enzymes of the prenylation pathway [87, 88].

Another type of lipid-related post-translational modification is termed acylation, which is the process of adding fatty acids to amino acids. Protein acylation is tightly regulated by histone acetyltransferases (HATs) and deacetylases (HDACs), and modulates various cellular functions such as cell proliferation, differentiation, and migration [89]. HATs have been reported to modulate cancer in two ways depending on the site of acetylation and type of cancer-one pro-tumorigenic and the other tumour-suppressive [90]. For instance, histone hyperacetylation was reported in liver cancer cells [91] whereas deficiency in acetylation was observed in prostate cancer patients [92]. In gastrointestinal carcinomas, decreased histone acetylation is significantly associated with severity of tumour invasion and metastasis [93]. Moreover, Kang and colleagues [94] demonstrated that curcumin-induced histone hypoacetylation triggers caspase-3-dependent apoptosis and promotes neuron differentiation of neural progenitor cells in brain cancer. The role of HDACs in cancer was also demonstrated in several cancers such as cervical [95], colon [96], and gastric cancer [97]. Similar to HATs, HDACs also have a dual function in cancer regulation. For example, loss of HDAC1 in teratomas increased apoptosis and induced cell arrest, albeit no change in tumour size [98]. Similarly, increase in cellular differentiation and apoptosis was observed when HDAC2 expression was ablated in colorectal cancer cells [95]. In contrast, knockdown of HDAC6 promoted migration and tube formation in HUVEC cells *in vitro* [99].

The modification of proteins by lipids is also important for cellular localization and transport [100]. For example, attachment of GPI to proteins triggers translocation to the outer leaflet of the plasma membrane, which is important for signal transduction events [101]. Therefore, the knowledge of different types of lipid-based post-translational modification of proteins is useful to dissect the causal effects of these modifications in the context of cancer biology.

4. Lipids and autophagy in cancer

The recycling and circulation of lipids within a cell is regulated by lysosomes, a membrane enclosed organelle containing hydrolytic enzymes [102]. In recent years, there have been emerging studies indicating the importance of lysosomal-mediated degradation, a process termed autophagy, in maintaining cellular lipid homeostasis in various tissues [103]. Autophagy is essential for cell survival in the event of nutrient deprivation, where intracellular proteins and organelles are targeted to the lysosome for degradation as an alternative source of recycled energy [104]. There are three commonly described autophagy processes: autophagy (also referred as macroautophagy) [105], microautophagy [106], and chaperone-mediated autophagy [107]. Dysregulation in autophagy is associated with a wide array of diseases such as metabolic, cardiovascular, and neurodegenerative diseases, ageing and cancer [108]. In addition to its role in starvation responses, growth and differentiation, and the clearance of dysfunctional/damaged cytoplasmic protein and organelles, autophagy has also been reported in tumour regulation in cancer [109].

The relationship between lipids and autophagy is of particular interest as autophagy has been widely established to have a role in cancer, albeit a complicated one. Some reports have stated that early in tumorigenesis, autophagy may act as a tumour suppressor mechanism (reviewed in [110, 111]). Beclin-1, the mammalian ortholog of yeast autophagy-related gene 6 (Atg6), has been widely accepted as a candidate for tumour suppression. Allelic deletion of Beclin-1 [112] and reduced protein expression [113] was observed in ovarian, breast, and prostate cancers. Beclin $1^{+/-}$ heterozygous mutant mice had reduced autophagic activity and spontaneous tumour development [114], indicating the importance of Beclin-1 in the causal effect of autophagy and tumour growth. However, as cancer progresses, autophagy becomes essential to overcome oxidative and metabolic stressors in the cell, hence improving cancer cell survival and progression [115]. For example, human cancer cells expressing the Ras oncogene are able to upregulate autophagy to support tumorigenesis and tumour cell survival under starvation conditions [116]. As autophagy can facilitate or suppress the development of cancer, targeting this facet as a cancer therapy should focus on both the regulation and inhibition of autophagy at the appropriate stages. It still nevertheless holds potential as a primary target or co-target as multiple studies have shown that inhibition of autophagy enhanced therapeutic effects against cancer in myeloma, breast, colon, and prostate cancer [117].

Lipids and lipid enzymes have indispensable roles in the autophagic process and can influence autophagy at various stages [118, 119]. For instance, the mTOR complex is an important negative regulator of autophagy and lipids such as phosphatidylinositol 3-phosphate (PI3P), diacylglycerol, and phosphatidic acids interfere with mTOR downstream signalling by acting independently to promote autophagy [118, 120]. During later stages of autophagy, cellular materials targeted for degradation are signalled to autophagosomes. Lipid droplets and the lipid enzyme phospholipase D have been postulated to regulate autophagosomes biogenesis as well as positively modulate autophagy *in vivo* and *in vitro* [121, 122]. Furthermore, Seo and colleagues shown that upon starvation, SREBPs can directly activate genes related to autophagy and are required for autophagosome formation and association with lipid droplets.

5. Lipids in angiogenesis and lymphangiogenesis

Classic characteristics of malignant tumours are their augmented proliferative and invasive properties. In order for cancer cells to sustain these enhanced growth requirements as well as expansion into other tissues, they have been shown to induce angiogenesis for oxygen and nutrient supply [123]. Tumour vasculature is also useful for the clearance of metabolic end products such as lactic acid whose accumulation may be toxic to the tumour cells. New capillary formation into tumours can be stimulated by growth factors such as vascular endothelial growth factor (VEGF) and fibroblast growth factor (bFGF) [124, 125]. In normal healthy cells, VEGF functions by creating new blood vessels during embryonic development and wound healing [126]. The tumour microenvironment is made up of a variety of cell types that are normal or quiescent. As a tumour expands in size, nutrient deprivation and hypoxia occurs. This triggers the production of VEGF and cytokines by the tumour into its surrounding microenvironment [127], thereby initiating the proliferation of endothelial cells which allows tumours to develop and grow exponentially. Although this vasculature initiation may provide the tumour with more oxygen and nutrients, the eventual outcome is not ideal. VEGF-induced formation of tumour vasculature are irregularly shaped, leaky, and often functionally abnormal [124]. The leaky nature of these tumour vasculature triggers the recruitment of platelets, which subsequently releases angiogenic stimulatory factors into the microenvironment to further promote angiogenesis [128]. Other than dissemination through blood vessels, tumour cells can also exploit the lymphatic vessel pathway for invasion into other tissues, hence promoting metastasis [129]. In particular, VEGF-C is the main mediator of lymphangiogenesis and lymph node metastasis [130].

The importance of lipids in tumour angiogenesis is highlighted in studies related to the bioactive sphingolipid derivative S1P. The function of S1P is comparable to growth factors VEGF and bFGF, where its secretion stimulates angiogenesis [131] and vascular maturation [132]. Interactions between S1P and these proangiogenic growth factors have also been reported and may provide a collective effect in promoting development of the vascular network [133]. S1P expression is upregulated in various tumours such as lung [134] and colorectal cancer [135]. Cancer cells are able to secrete S1P into their microenvironment to induce both angiogenesis and lymphangiogenesis [136, 137]; via binding of S1P receptors [138], thereby facilitating tumour spread. Furthermore, *in vitro* analysis revealed that high levels of S1P are associated with increased migration and tube formation in co-cultured vascular or lymphatic endothelial cells [139]. Angiogenic and lymphatic metastasis is also stimulated by the secretion of prostaglandins, a group of lipid compounds enzymatically derived from fatty acids [140]. In particular, PGE(2) in breast cancer is able to bind GPCRs and induce angiogenic regulatory genes for proliferation, tube formation and subsequently metastasis [141]. This was also true in prostate cancer where PGE(2) activates angiogenesis via the prostanoids EP2 and EP4 pathways to increase production of urokinase-type plasminogen and vascular endothelial growth factors to alter prostate cancer cell motility [142].

Lipid metabolism has also been implicated in angiogenesis. SREBP1 expression is elevated in newly formed vasculature [143]. In response to VEGF signals, endothelial cells activate SREBP1 and SREBP2 to trigger proliferation, migration, and vascular formation [144]. Vice versa, inhibition of SREBP1 resulted in reduced production of pro-angiogenic factors [143]. Metastasis is one of the main causes of mortality in human cancers. Since angiogenesis and lymphangiogenesis provide a platform for tumours to acquire nutrients and metastasise, understanding the role of lipids in endothelial cell metabolism may be useful as a target for cancer therapy and drug resistance [145, 146].

6. Concluding remarks

Lipid metabolism and signalling are now widely accepted as major players in cancer biology. Targeting components such as enzymes, bioactive lipids, and receptors, all of which are important for maintaining lipid homeostasis, metabolism and signalling, have been shown to reduce cancer cell proliferation and metastasis. This can be achieved through various means such as modifying the function of enzymes involved in biosynthesis and metabolism of lipids, altering the structure, composition and localisation of bioactive lipids and lipid rafts, or through disruption of lipid-mediated tumour-stromal crosstalk in the tumour microenvironment, and by promoting apoptosis of cancer cells. Considering the central role of lipids in cancer, these strategies are encouraging for the treatment and cure against cancer.

Acknowledgements

This work was supported by the Early Career Scheme GRF-HKBU-22103017-ECS of the Hong Kong Research Grants Council.

Author details

Jin Yan Lim and Hiu Yee Kwan*
Centre for Cancer and Inflammation Research, School of Chinese Medicine, Hong Kong Baptist University, Hong Kong

*Address all correspondence to: hykwan@hkbu.edu.hk

IntechOpen

References

[1] Beloribi-Djefaflia S, Vasseur S, Guillaumond F. Lipid metabolic reprogramming in cancer cells. Oncogene. 2016;5:e189

[2] Vander Heiden MG, Cantley LC, Thompson CB. Understanding the Warburg effect: The metabolic requirements of cell proliferation. Science. 2009;324(5930):1029-1033

[3] Liberti MV, Locasale JW. The Warburg effect: How does it benefit cancer cells? Trends in Biochemical Sciences. 2016;41(3):211-218

[4] Baenke F, Peck B, Miess H, Schulze A. Hooked on fat: The role of lipid synthesis in cancer metabolism and tumour development. Disease Models & Mechanisms. 2013;6(6):1353-1363

[5] Menendez JA, Lupu R. Fatty acid synthase and the lipogenic phenotype in cancer pathogenesis. Nature Reviews. Cancer. 2007;7(10):763-777

[6] Li J, Cheng J-X. Direct visualization of de novo lipogenesis in single living cells. Scientific Reports. 2014;4:6807

[7] Rysman E et al. De novo lipogenesis protects cancer cells from free radicals and chemotherapeutics by promoting membrane lipid saturation. Cancer Research. 2010;70(20):8117-8126

[8] Kannar R, Lyon I, Baker N. Dietary control of lipogenesis in vivo in host tissues and tumors of mice bearing Ehrlich ascites carcinoma. Cancer Research. 1980;40(12):4606-4611

[9] Ookhtens M, Kannan R, Lyon I, Baker N. Liver and adipose tissue contributions to newly formed fatty acids in an ascites tumor. The American Journal of Physiology. 1984;247(1 Pt 2): R146-R153

[10] Chajes V, Cambot M, Moreau K, Lenoir GM, Joulin V. Acetyl-CoA carboxylase alpha is essential to breast cancer cell survival. Cancer Research. 2006;66(10):5287-5294

[11] Zaidi N, Swinnen JV, Smans K. ATP-citrate lyase: A key player in cancer metabolism. Cancer Research. 2012;72(15):3709-3714

[12] Flavin R, Peluso S, Nguyen PL, Loda M. Fatty acid synthase as a potential therapeutic target in cancer. Future Oncology. 2010;6(4):551-562

[13] Migita T et al. ATP citrate lyase: Activation and therapeutic implications in non-small cell lung cancer. Cancer Research. 2008;68(20):8547-8554

[14] Buckley D et al. Fatty acid synthase—Modern tumor cell biology insights into a classical oncology target. Pharmacology & Therapeutics. 2017;177:23-31

[15] Svensson RU et al. Inhibition of acetyl-CoA carboxylase suppresses fatty acid synthesis and tumor growth of non-small-cell lung cancer in preclinical models. Nature Medicine. 2016;22:1108

[16] Hatzivassiliou G et al. ATP citrate lyase inhibition can suppress tumor cell growth. Cancer Cell. 2005;8(4):311-321

[17] Horton JD, Goldstein JL, Brown MS. SREBPs: Activators of the complete program of cholesterol and fatty acid synthesis in the liver. The Journal of Clinical Investigation. 2002;109(9):1125-1131

[18] Peterson TR et al. mTOR. Complex 1 regulates lipin 1 localization to control the SREBP pathway. Cell. 2011;146(3):408-420

[19] Guertin DA, Sabatini DM. Defining the role of mTOR in cancer. Cancer Cell. 2007;**12**(1):9-22

[20] Saxton RA, Sabatini DM. mTOR signaling in growth, metabolism, and disease. Cell. 2017;**168**(6):960-976

[21] Porstmann T et al. SREBP activity is regulated by mTORC1 and contributes to Akt-dependent cell growth. Cell Metabolism. 2008;**8**(3):224-236

[22] Duvel K et al. Activation of a metabolic gene regulatory network downstream of mTOR complex 1. Molecular Cell. 2010;**39**(2):171-183

[23] Holz MK, Ballif BA, Gygi SP, Blenis J. mTOR and S6K1 mediate assembly of the translation preinitiation complex through dynamic protein interchange and ordered phosphorylation events. Cell. 2005;**123**(4):569-580

[24] Ben-Sahra I, Howell JJ, Asara JM, Manning BD. Stimulation of de novo pyrimidine synthesis by growth signaling through mTOR and S6K1. Science. 2013;**339**(6125):1323-1328

[25] Ben-Sahra I, Hoxhaj G, Ricoult SJH, Asara JM, Manning BD. mTORC1 induces purine synthesis through control of the mitochondrial tetrahydrofolate cycle. Science. 2016;**351**(6274):728-733

[26] Eggens I, Bäckman L, Jakobsson A, Valtersson C. The lipid composition of highly differentiated human hepatomas, with special reference to fatty acids. British Journal of Experimental Pathology. 1988;**69**(5):671-683

[27] Yates AJ, Thompson DK, Boesel CP, Albrightson C, Hart RW. Lipid composition of human neural tumors. Journal of Lipid Research. 1979;**20**(4):428-436

[28] Portoukalian J, Zwingelstein G, Dore J. Lipid composition of human malignant melanoma tumors at various levels of malignant growth. European Journal of Biochemistry. 1979;**94**(1):19-23

[29] Gray GM. The lipid composition of tumour cells. The Biochemical Journal. 1963;**86**(2):350-357

[30] He M, Guo S, Li Z. In situ characterizing membrane lipid phenotype of breast cancer cells using mass spectrometry profiling. Scientific Reports. 2015;**5**:11298

[31] Guri Y et al. mTORC2 promotes tumorigenesis via lipid synthesis. Cancer Cell. 2017;**32**(6):807-823.e12

[32] Pyne NJ, Pyne S. Sphingosine 1-phosphate and cancer. Nature Reviews. Cancer. 2010;**10**:489

[33] Ogretmen B. Sphingolipid metabolism in cancer signalling and therapy. Nature Reviews. Cancer. 2018;**18**(1):33-50

[34] Liśkiewicz AD et al. Long-term high fat ketogenic diet promotes renal tumor growth in a rat model of tuberous sclerosis. Scientific Reports. 2016;**6**:21807

[35] Xia S et al. Prevention of dietary-fat-fueled ketogenesis attenuates BRAF V600E tumor growth. Cell Metabolism. 2017;**25**(2):358-373

[36] Chen M et al. An aberrant SREBP-dependent lipogenic program promotes metastatic prostate cancer. Nature Genetics. 2018;**50**(2):206-218

[37] Wang B et al. Phospholipid remodeling and cholesterol availability regulate intestinal stemness and tumorigenesis. Cell Stem Cell. 2018;**22**(2):206-220.e4

[38] Pascual G et al. Targeting metastasis-initiating cells through

the fatty acid receptor CD36. Nature. 2017;**541**(7635):41-45

[39] Mebratu Y, Tesfaigzi Y. How ERK1/2 activation controls cell proliferation and cell death is subcellular localization the answer? Cell Cycle. 2009;**8**(8):1168-1175

[40] Binker-Cosen MJ, Richards D, Oliver B, Gaisano HY, Binker MG, Cosen-Binker LI. Palmitic acid increases invasiveness of pancreatic cancer cells AsPC-1 through TLR4/ROS/ NF-κB/MMP-9 signaling pathway. Biochemical and Biophysical Research Communications. 2017;**484**(1):152-158

[41] Kwan HY et al. Subcutaneous adipocytes promote melanoma cell growth by activating the Akt signaling pathway: Role of palmitic acid. The Journal of Biological Chemistry. 2014;**289**(44):30525-30537

[42] Zhang C, Yu H, Ni X, Shen S, Das UN. Growth inhibitory effect of polyunsaturated fatty acids (PUFAs) on colon cancer cells via their growth inhibitory metabolites and fatty acid composition changes. PLoS One. 2015;**10**(4):e0123256

[43] Mac Lennan M, Ma DWL. Role of dietary fatty acids in mammary gland development and breast cancer. Breast Cancer Research. 2010;**12**(5):211

[44] Theodoratou E et al. Dietary fatty acids and colorectal cancer: A case-control study. American Journal of Epidemiology. 2007;**166**(2):181-195

[45] Vaughan VC, Hassing M-R, Lewandowski PA. Marine polyunsaturated fatty acids and cancer therapy. British Journal of Cancer. 2013;**108**(3):486-492

[46] Li J et al. Dietary supplementation of α-linolenic acid induced conversion of n-3 LCPUFAs and reduced prostate cancer growth in a mouse model. Lipids in Health and Disease. 2017;**16**(1):136

[47] Azrad M et al. Prostatic alpha-linolenic acid (ALA) is positively associated with aggressive prostate cancer: A relationship which may depend on genetic variation in ALA metabolism. PLoS One. 2012;**7**(12):e53104

[48] Nabavi SF et al. Omega-3 polyunsaturated fatty acids and cancer: Lessons learned from clinical trials. Cancer Metastasis Reviews. 2015;**34**(3):359-380

[49] Murff HJ et al. Dietary polyunsaturated fatty acids and breast cancer risk in Chinese women: A prospective cohort study. International Journal of Cancer. 2011;**128**(6):1434-1441

[50] Rhodes LE et al. Effect of eicosapentaenoic acid, an omega-3 polyunsaturated fatty acid, on UVR-related cancer risk in humans. An assessment of early genotoxic markers. Carcinogenesis. 2003;**24**(5):919-925

[51] Cockbain AJ et al. Anticolorectal cancer activity of the omega-3 polyunsaturated fatty acid eicosapentaenoic acid. Gut. 2014;**63**(11):1760 LP-1761768

[52] Pappalardo G, Almeida A, Ravasco P. Eicosapentaenoic acid in cancer improves body composition and modulates metabolism. Nutrition. 2015;**31**(4):549-555

[53] Newell M, Baker K, Postovit LM, Field CJ. A critical review on the effect of docosahexaenoic acid (DHA) on cancer cell cycle progression. International Journal of Molecular Sciences. 2017;**18**(8):1784

[54] Park M, Kim H. Anti-cancer mechanism of docosahexaenoic acid in pancreatic carcinogenesis: A mini-review. Journal of Cancer Prevention. 2017;**22**(1):1-5

[55] Arab A, Akbarian SA, Ghiyasvand R, Miraghajani M. The effects of conjugated linoleic acids on breast cancer: A systematic review. Advanced Biomedical Research. 2016;**5**:115

[56] Zhou Y, Wang T, Zhai S, Li W, Meng Q. Linoleic acid and breast cancer risk: A meta-analysis. Public Health Nutrition. 2016;**19**(8):1457-1463

[57] Ghosh J, Myers CE. Arachidonic acid stimulates prostate cancer cell growth: Critical role of 5-lipoxygenase. Biochemical and Biophysical Research Communications. 1997;**235**(2):418-423

[58] Hughes-Fulford M, Li C-F, Boonyaratanakornkit J, Sayyah S. Arachidonic acid activates phosphatidylinositol 3-kinase signaling and induces gene expression in prostate cancer. Cancer Research. 2006;**66**(3):1427-1433

[59] Sakai M et al. Arachidonic acid and cancer risk: A systematic review of observational studies. BMC Cancer. 2012;**12**:606

[60] Panupinthu N, Lee HY, Mills GB. Lysophosphatidic acid production and action: Critical new players in breast cancer initiation and progression. British Journal of Cancer. 2010;**102**(6):941-946

[61] Benesch MGK, Ko YM, McMullen TPW, Brindley DN. Autotaxin in the crosshairs: Taking aim at cancer and other inflammatory conditions. FEBS Letters. 2014;**588**(16):2712-2727

[62] Nam SW, Clair T, Campo CK, Lee HY, Liotta LA, Stracke ML. Autotaxin (ATX), a potent tumor motogen, augments invasive and metastatic potential of ras-transformed cells. Oncogene. 2000;**19**(2):241-247

[63] Kishi Y et al. Autotaxin is overexpressed in glioblastoma multiforme and contributes to cell motility of glioblastoma by converting lysophosphatidylcholine to lysophosphatidic acid. The Journal of Biological Chemistry. 2006;**281**(25):17492-17500

[64] Nouh MAAM et al. Expression of autotaxin and acylglycerol kinase in prostate cancer: Association with cancer development and progression. Cancer Science. 2009;**100**(9):1631-1638

[65] Yang SY et al. Expression of autotaxin (NPP-2) is closely linked to invasiveness of breast cancer cells. Clinical & Experimental Metastasis. 2002;**19**(7):603-608

[66] Park SY et al. Lysophosphatidic acid augments human hepatocellular carcinoma cell invasion through LPA1 receptor and MMP-9 expression. Oncogene. 2010;**30**:1351

[67] Hannun YA, Obeid LM. Sphingolipids and their metabolism in physiology and disease. Nature Reviews. Molecular Cell Biology. 2018;**19**(3):175-191

[68] Ponnusamy S et al. Sphingolipids and cancer: Ceramide and sphingosine-1-phosphate in the regulation of cell death and drug resistance. Future Oncology. 2010;**6**(10):1603-1624

[69] Pettus BJ, Chalfant CE, Hannun YA. Ceramide in apoptosis: An overview and current perspectives. Biochimica et Biophysica Acta. 2002;**1585**(2-3):114-125

[70] Chang K-T, Anishkin A, Patwardhan GA, Beverly LJ, Siskind LJ, Colombini M. Ceramide channels: Destabilization by Bcl-xL and role in apoptosis. Biochimica et Biophysica Acta. 2015;**1848**(10):2374-2384

[71] Hla T. Physiological and pathological actions of sphingosine 1-phosphate. Seminars in Cell & Developmental Biology. 2004;**15**(5):513-520

[72] Maceyka M, Payne SG, Milstien S, Spiegel S. Sphingosine kinase, sphingosine-1-phosphate, and apoptosis. Biochimica et Biophysica Acta. 2002;**1585**(2-3):193-201

[73] Rosen H, Goetzl EJ. Sphingosine 1-phosphate and its receptors: An autocrine and paracrine network. Nature Reviews. Immunology. 2005;**5**(7):560-570

[74] Nakanishi M, Rosenberg DW. Multifaceted roles of PGE2 in inflammation and cancer. Seminars in Immunopathology. 2013;**35**(2):123-137

[75] Kawamori T, Uchiya N, Sugimura T, Wakabayashi K. Enhancement of colon carcinogenesis by prostaglandin E2 administration. Carcinogenesis. 2003;**24**(5):985-990

[76] Wang D, Buchanan FG, Wang H, Dey SK, DuBois RN. Prostaglandin E2 enhances intestinal adenoma growth via activation of the Ras-mitogen-activated protein kinase cascade. Cancer Research. 2005;**65**(5):1822-1829

[77] Chandramouli A et al. MicroRNA-101 (miR-101) post-transcriptionally regulates the expression of EP4 receptor in colon cancers. Cancer Biology & Therapy. 2012;**13**(3):175-183

[78] Doherty GA et al. Proneoplastic effects of PGE2 mediated by EP4 receptor in colorectal cancer. BMC Cancer. 2009;**9**:207

[79] Jin J et al. Prostanoid EP1 receptor as the target of (−)-epigallocatechin-3-gallate in suppressing hepatocellular carcinoma cells in vitro. Acta Pharmacologica Sinica. 2012;**33**(5):701-709

[80] Amano H et al. Roles of a prostaglandin E-type receptor, EP3, in upregulation of matrix metalloproteinase-9 and vascular endothelial growth factor during enhancement of tumor metastasis. Cancer Science. 2009;**100**(12):2318-2324

[81] Krueger KE, Srivastava S. Posttranslational protein modifications: Current implications for cancer detection, prevention, and therapeutics. Molecular & Cellular Proteomics. 2006;**5**(10):1799-1810

[82] Wang M, Casey PJ. Protein prenylation: Unique fats make their mark on biology. Nature Reviews. Molecular Cell Biology. 2016;**17**(2):110-122

[83] Sebti SM. Protein farnesylation: Implications for normal physiology, malignant transformation, and cancer therapy. Cancer Cell. 2005;**7**(4):297-300

[84] Schubbert S, Shannon K, Bollag G. Hyperactive Ras in developmental disorders and cancer. Nature Reviews. Cancer. 2007;**7**(4):295-308

[85] Giehl K. Oncogenic Ras in tumour progression and metastasis. Biological Chemistry. 2005;**386**(3):193-205

[86] Shamma A et al. Rb regulates DNA damage response and cellular senescence through E2F-dependent suppression of N-ras isoprenylation. Cancer Cell. 2009;**15**(4):255-269

[87] Kloog Y, Cox AD. Prenyl-binding domains: Potential targets for Ras inhibitors and anti-cancer drugs. Seminars in Cancer Biology. 2004;**14**(4):253-261

[88] Nguyen UTT, Goody RS, Alexandrov K. Understanding and exploiting protein prenyltransferases. ChemBioChem. 2010;**11**(9):1194-1201

[89] Liu N, Li S, Wu N, Cho K-S. Acetylation and deacetylation in cancer stem-like cells. Oncotarget. 2017;**8**(51):89315-89325

[90] DiCerbo V, Schneider R. Cancers with wrong HATs: The impact of acetylation. Briefings in Functional Genomics. 2013;**12**(3):231-243

[91] Bai X et al. Overexpression of myocyte enhancer factor 2 and histone hyperacetylation in hepatocellular carcinoma. Journal of Cancer Research and Clinical Oncology. 2008;**134**(1):83-91

[92] Cang S et al. Deficient histone acetylation and excessive deacetylase activity as epigenomic marks of prostate cancer cells. International Journal of Oncology. 2009;**35**(6):1417-1422

[93] Yasui W, Oue N, Ono S, Mitani Y, Ito R, Nakayama H. Histone acetylation and gastrointestinal carcinogenesis. Annals of the New York Academy of Sciences. 2003;**983**:220-231

[94] Kang S-K, Cha S-H, Jeon H-G. Curcumin-induced histone hypoacetylation enhances caspase-3-dependent glioma cell death and neurogenesis of neural progenitor cells. Stem Cells and Development. 2006;**15**(2):165-174

[95] Huang BH et al. Inhibition of histone deacetylase 2 increases apoptosis and p21Cip1/WAF1 expression, independent of histone deacetylase 1. Cell Death and Differentiation. 2005;**12**(4):395-404

[96] Wilson AJ et al. Histone deacetylase 3 (HDAC3) and other class I HDACs regulate colon cell maturation and p21 expression and are deregulated in human colon cancer. The Journal of Biological Chemistry. 2006;**281**(19):13548-13558

[97] Song J et al. Increased expression of histone deacetylase 2 is found in human gastric cancer. APMIS. 2005;**113**(4):264-268

[98] Lagger S et al. Crucial function of histone deacetylase 1 for differentiation of teratomas in mice and humans. The EMBO Journal. 2010;**29**(23):3992-4007

[99] Lv Z et al. Downregulation of HDAC6 promotes angiogenesis in hepatocellular carcinoma cells and predicts poor prognosis in liver transplantation patients. Molecular Carcinogenesis. 2016;**55**(5):1024-1033

[100] Sezgin E, Levental I, Mayor S, Eggeling C. The mystery of membrane organization: Composition, regulation and roles of lipid rafts. Nature Reviews. Molecular Cell Biology. 2017;**18**(6):361-374

[101] Paulick MG, Bertozzi CR. The glycosylphosphatidylinositol anchor: A complex membrane-anchoring structure for proteins. Biochemistry. 2008;**47**(27):6991-7000

[102] Cooper GM. Lysosomes. In: The Cell: A Molecular Approach. 2nd ed. Sunderland (MA): Sinauer Associates; 2000

[103] Singh R et al. Autophagy regulates lipid metabolism. Nature. 2009;**458**(7242):1131-1135

[104] Mizushima N. Autophagy: Process and function. Genes & Development. 2007;**21**(22):2861-2873

[105] Yu L, Chen Y, Tooze SA. Autophagy pathway: Cellular and molecular mechanisms. Autophagy. 2018;**14**(2):207-215

[106] Li W, Li J, Bao J. Microautophagy: Lesser-known self-eating. Cellular and Molecular Life Sciences. 2012;**69**(7):1125-1136

[107] Kaushik S, Cuervo AM. Chaperone-mediated autophagy: A unique way to enter the lysosome world. Trends in Cell Biology. 2012;**22**(8):407-417

[108] Levine B, Kroemer G. Autophagy in the pathogenesis of disease. Cell. 2008;**132**(1):27-42

[109] Mizushima N. The pleiotropic role of autophagy: From protein metabolism to bactericide. Cell Death and Differentiation. 2005;**12**(Suppl 2): 1535-1541

[110] Avalos Y, Canales J, Bravo-Sagua R, Criollo A, Lavandero S, Quest AFG. Tumor suppression and promotion by autophagy. BioMed Research International. 2014;**2014**:603980

[111] Gozuacik D, Kimchi A. Autophagy as a cell death and tumor suppressor mechanism. Oncogene. 2004;**23**:2891

[112] Aita VM et al. Cloning and genomic organization of beclin 1, a candidate tumor suppressor gene on chromosome 17q21. Genomics. 1999;**59**(1):59-65

[113] Liang XH et al. Induction of autophagy and inhibition of tumorigenesis by beclin 1. Nature. 1999;**402**:672

[114] Qu X et al. Promotion of tumorigenesis by heterozygous disruption of the beclin 1 autophagy gene. The Journal of Clinical Investigation. 2003;**112**(12):1809-1820

[115] White E. The role for autophagy in cancer. The Journal of Clinical Investigation. 2015;**125**(1):42-46

[116] Guo JY et al. Activated Ras requires autophagy to maintain oxidative metabolism and tumorigenesis. Genes & Development. 2011;**25**(5):460-470

[117] Chen N, Karantza V. Autophagy as a therapeutic target in cancer. Cancer Biology & Therapy. 2011;**11**(2):157-168

[118] Dall'Armi C, Devereaux KA, Di Paolo G. The role of lipids in the control of autophagy. Current Biology. 2013;**23**(1):R33-R45

[119] Jaishy B, Abel ED. Lipids, lysosomes, and autophagy. Journal of Lipid Research. 2016;**57**(9):1619-1635

[120] Zoncu R, Efeyan A, Sabatini DM. mTOR: From growth signal integration to cancer, diabetes and ageing. Nature Reviews. Molecular Cell Biology. 2011;**12**(1):21-35

[121] Dall'Armi C et al. The phospholipase D1 pathway modulates macroautophagy. Nature Communications. 2010;**1**:142

[122] Shpilka T et al. Lipid droplets and their component triglycerides and steryl esters regulate autophagosome biogenesis. The EMBO Journal. 2015;**34**(16):2117-2131

[123] Folkman J. Role of angiogenesis in tumor growth and metastasis. Seminars in Oncology. 2002;**29**(6 Suppl 16):15-18

[124] Carmeliet P. VEGF as a key mediator of angiogenesis in cancer. Oncology. 2005;**69**(Suppl 3):4-10

[125] Ucuzian AA, Gassman AA, East AT, Greisler HP. Molecular mediators of angiogenesis. Journal of Burn Care & Research. 2010;**31**(1):158

[126] Shibuya M. Vascular endothelial growth factor (VEGF) and its receptor (VEGFR) signaling in angiogenesis: A crucial target for anti- and pro-angiogenic therapies. Genes & Cancer. 2011;**2**(12):1097-1105

[127] Folkman J, Hanahan D. Switch to the angiogenic phenotype during tumorigenesis. Princess Takamatsu Symposia. 1991;**22**:339-347

[128] Weis SM, Cheresh DA. Pathophysiological consequences of

VEGF-induced vascular permeability. Nature. 2005;**437**(7058):497-504

[129] Paduch R. The role of lymphangiogenesis and angiogenesis in tumor metastasis. Cellular Oncology. 2016;**39**(5):397-410

[130] Michael MS, Pepper S, Tille J-C, Nisato R. Lymphangiogenesis and tumor metastasis. Cell and Tissue Research. 2003;**314**(1):167-177

[131] Lee OH et al. Sphingosine 1-phosphate induces angiogenesis: Its angiogenic action and signaling mechanism in human umbilical vein endothelial cells. Biochemical and Biophysical Research Communications. 1999;**264**(3):743-750

[132] Liu Y et al. Edg-1, the G protein-coupled receptor for sphingosine-1-phosphate, is essential for vascular maturation. The Journal of Clinical Investigation. 2000;**106**(8):951-961

[133] Spiegel S, Milstien S. Sphingosine-1-phosphate: An enigmatic signalling lipid. Nature Reviews. Molecular Cell Biology. 2003;**4**(5):397-407

[134] Johnson KR et al. Immunohistochemical distribution of sphingosine kinase 1 in normal and tumor lung tissue. The Journal of Histochemistry and Cytochemistry. 2005;**53**(9):1159-1166

[135] Kawamori T et al. Role for sphingosine kinase 1 in colon carcinogenesis. The FASEB Journal. 2009;**23**(2):405-414

[136] Nagahashi M et al. Sphingosine-1-phosphate produced by sphingosine kinase 1 promotes breast cancer progression by stimulating angiogenesis and lymphangiogenesis. Cancer Research. 2012;**72**(3):726-735

[137] Visentin B et al. Validation of an anti-sphingosine-1-phosphate antibody as a potential therapeutic in reducing growth, invasion, and angiogenesis in multiple tumor lineages. Cancer Cell. 2006;**9**(3):225-238

[138] English D, Brindley DN, Spiegel S, Garcia JGN. Lipid mediators of angiogenesis and the signalling pathways they initiate. Biochimica et Biophysica Acta. 2002;**1582**(1-3):228-239

[139] Anelli V, Gault CR, Snider AJ, Obeid LM. Role of sphingosine kinase-1 in paracrine/transcellular angiogenesis and lymphangiogenesis in vitro. The FASEB Journal. 2010;**24**(8):2727-2738

[140] Karnezis T et al. VEGF-D promotes tumor metastasis by regulating prostaglandins produced by the collecting lymphatic endothelium. Cancer Cell. 2012;**21**(2):181-195

[141] Chang S-H et al. Role of prostaglandin E2-dependent angiogenic switch in cyclooxygenase 2-induced breast cancer progression. Proceedings of the National Academy of Sciences of the United States of America. 2004;**101**(2):591-596

[142] Jain S, Chakraborty G, Raja R, Kale S, Kundu GC. Prostaglandin E2 regulates tumor angiogenesis in prostate cancer. Cancer Research. 2008;**68**(19):7750-7759

[143] Min Y, Rui-Hai Z, Melissa P, Lei Z, John S, Manuela M-G. Activation of sterol regulatory element-binding proteins (SREBPs) is critical in IL-8-induced angiogenesis. Journal of Leukocyte Biology. 2006;**80**(3):608-620

[144] Zhou R-H, Yao M, Lee T-S, Zhu Y, Martins-Green M, Shyy JY-J. Vascular endothelial growth factor activation of sterol regulatory element binding protein: A potential role in angiogenesis. Circulation Research. 2004;**95**(5):471-478

[145] Iwamoto H et al. Cancer lipid metabolism confers antiangiogenic drug resistance. Cell Metabolism. 2018;**28**(1):104-117

[146] Rohlenova K, Veys K, Miranda-Santos I, DeBock K, Carmeliet P. Endothelial cell metabolism in health and disease. Trends in Cell Biology. 2018;**28**(3):224-236

Chapter 2

Analytical Tools for Lipid Assessment in Biological Assays

Banny Silva Barbosa Correia, Raquel Susana Torrinhas,
William Yutaka Ohashi and Ljubica Tasic

Abstract

Lipids are heterogeneous biological molecules with many important roles. In human body, lipids can be energy substrates, steroid hormones, inflammatory lipid mediators, transporters, and feature as structural cell and organelle membrane elements. At the cell membrane, lipids influence the distribution of surface proteins and, in part, protein signaling and, consequently, the activation of transcriptional factors. One of the best explored relationships in chemistry and science is the structure/activity one. Therefore, if the composition of a mixture is discovered and the structure of its components is known, a task of proposing relationship among all components and their activity would be closer to understanding. There are many powerful and advantageous analytical and bioanalytical tools available for the study of lipids, but all show at least some limitations. Knowing the advantages/disadvantages of each technique is essential for choosing the most appropriate one for the analysis as to answer a scientific question about lipid composition and role within a biological model. Often, inexperience and little familiarity with the cited analytical resources may limit the validity of the obtained results. Our chapter aims to present and discuss different tools available for the study of lipids and their main applications in biological assays.

Keywords: lipids, bioanalytical tools, gas chromatography, mass spectrometry, nuclear magnetic resonance

1. Introduction

Lipids are a very heterogeneous group of biological molecules. Some of the most studied lipids are built from the fatty acids (FAs) or isoprenyl groups. FAs are carboxylic acids composed by an even number of carbon atoms connected by single or double bonds with a methyl group end. FAs can be classified into very long (>20 carbons), long (14–20 carbons), medium (6–12 carbons), and short (up to 6 carbons)-chain FAs, as well as saturated (no double chains), monounsaturated (1 double bond), and polyunsaturated (PUFAs, >1 double bond) FAs. Furthermore, unsaturated fatty acids can receive its omega (n) assignment according to the first double-bond position from the end methyl group. Biosynthetically, endogenous FAs have been made from acetyl-CoA/malonyl-CoA [1–3].

FAs represent a class of lipids on their own and do not make part of all lipids [4]. Some lipids, which are not formed from FAs but are biosynthetically related to them, are the polyketides, formed from the acetyl units. Other unsaponifiable lipids

are built from isoprene units, molecules with five carbons with a branch structure
and alternated double bonds. Isoprenes have their biosynthesis in mevalonate (vegetables) or deoxyxylulose phosphate (animals) pathways. They can form sterols and
prenols [2]; some sterols can also have FAs in their structure [3].

Actually, lipids comprise eight main classes within different chemical characteristics: fatty acids (1), glycerolipids (2), glycerophospholipids (3), sphingolipids (4),
sterols (5), prenols (6), saccharolipids (7), and polyketides (8) (**Figure 1**) [3]. These
classes show a high diversity of molecules and are grouped into several subclasses.
Lipid classification based on their chemical information, described by the headgroup and the type of a linkage between the head group and aliphatic chains [5, 6]
is the most used among biochemists. Investigators have estimated the presence of
~180,000 lipid species in nature and ~40 common fatty acids as building blocks [7].
At the moment, 43,109 structurally distinct lipids are already registered at the Lipid
MAPS consortium.

The high diversity of lipids reflects their multiple biological functions and can
be attributed to the wide variety of their building blocks and numerous possible
permutations [6, 8]. In the human body, lipids serve as: substrates for the synthesis
of energy (9.3 kcal/g), steroid hormones, inflammatory lipid mediators, vitamins
or liposoluble vitamins transportation, and structural elements of cell and organelle membranes [9–11]. As a part of the cell membrane, lipids can influence the
distribution of surface proteins, protein signaling (as part of lipid rafts or as second
messengers), and consequently, the activation of transcriptional factors [12, 13].
This means that besides their recognized biological functions, lipids can influence
protein signaling and synthesis.

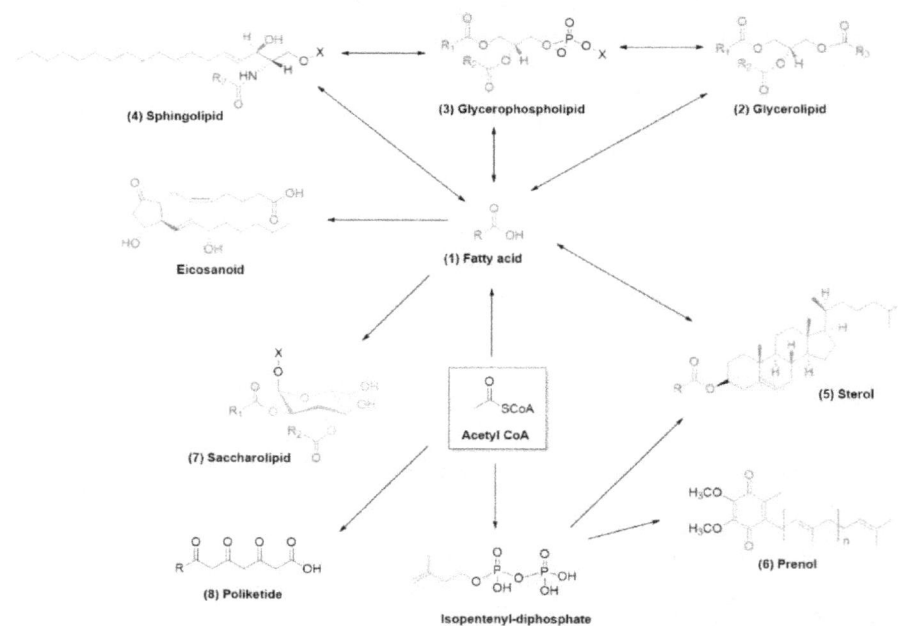

Figure 1.
*Biosynthetic lipid network. Acetyl-CoA: fatty acids-FAs (class 1) are synthetized, enabling the production
of other lipid classes: 2 (glycerolipids-GLs), 3 (glycerophospholipids-GPs), 4 (sphingolipids-SPs), and 7
(saccharolipids-SLs), as well the class of eicosanoids. Acetyl-CoA can also generate the class 8 (polyketides-
PKs) and isopentenyl diphosphate molecule, through mevalonate. On the other side, isoprenyl is used as starting
substrate for producing lipid classes 6 (prenols-PRs) and 5 (sterols-STs). Figure was inspired on Quhenberger
et al. [4].*

In a cell, lipids show different compositions, tens of thousands to hundreds of thousands of compounds, and concentrations from a mol/mg to nmol/mg of protein [5]. Facing the biological relevance of lipids, it is not surprising that the human organism has sophisticated machineries for the FA synthesis when its dietary supply flaws. Saturated and monounsaturated FAs can be endogenously generated from glucose and amino acids through enzymatic elongation (by adding units of two carbons) and desaturation (by forming new double bonds) reactions. However, a pitiful lack of the desaturating enzymes Δ-12 and Δ-15 desaturases preclude humans to add double bonds before the ninth carbon at the end of the methyl extremity for the synthesis of the polyunsaturated fatty acids (PUFAs) n-linoleic acid (C18:2 n-6, LA) and alpha-linolenic acid (C18:3 n-3, ALA). Consequently, LA and ALA are obtained exclusively from diet and, then, called as essentials. After ingestion, LA and ALA compete for sequential enzymatic processes of elongation and desaturation until their conversion into longer chain PUFAs: arachidonic acid (C20:4 n-6, ARA) from LA and eicosapentaenoic acid (C20:5 n-3, EPA) or docosahexaenoic acid (C22:6 n-3, DHA) from ALA [14].

ARA, EPA, and DHA have a high clinical interest once they influence the composition and steady-state of cell membranes. Also, they are precursors of the lipid mediators named eicosanoids involved in the activation of the inflammatory

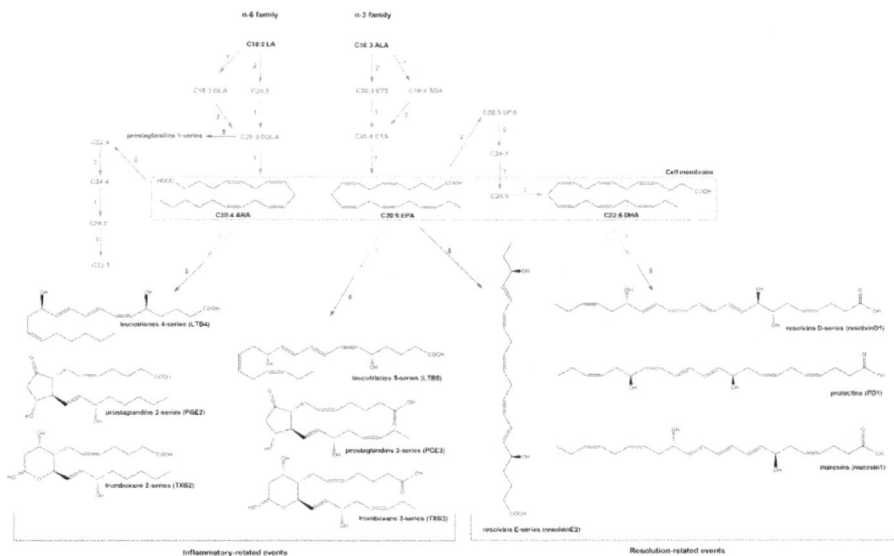

Figure 2.
Synthesis of lipid mediators from eicosapentaenoic (C20:2 n-6, EPA), docosahexaenoic (C22:6 n-3, DHA), and arachidonic (C20:4 n-6, ARA) acids. EPA, DHA, and ARA are previously synthetized from n-3 and n-6 fatty acid families in reactions mediated by enzymes: 1—desaturase, 2—elongase, 3—peroxisomal fatty acyl-CoA oxidase, 4—lipoxygenase (LOX), and 5—cyclooxygenase (COX). The cellular bioavailability of EPA decreases the production of ARA-produced eicosanoids, which include prostaglandins (PG)E2, thromboxane (TX) A2, and leukotriene (LT)B4. These eicosanoids have a higher pro-inflammatory potential than those contra parts produced from EPA (PGE5, TXA3, and LTB5) in promoting vasodilation and leukocyte chemotaxis and adhesion, events that stimulate the migration of neutrophils into the damaged tissue. As part of the neutrophil-monocyte sequence of inflammation, eicosanoids are no longer produced to initiate the synthesis of resolvins, protectins and maresins, lipid mediators from EPA and DHA. Other fatty acids shown are: linoleic acid (C18:2 n-6, LA), gamma-linolenic acid (C18:3 n-6, GLA), dihomo-gamma-linolenic acid (C20:3 n-6, DGLA), adrenic acid (C22:4 n-6), tetracosatetraenoic acid (C24:4 n-6), tetracosapentaenoic acid (C24:5 n-6), docosapentaenoic acid (C22:5 n-6), oleic acid (C18:1 n-9), octadecadienoic acid (C18:2 n-9), alpha-linolenic acid (C18:3 n-3, ALA), stearidonic acid (C18:4 n-3, SDA), eicosatrienoic acid (C20:3 n-3, ETE), eicosatetraenoic acid (C20:4 n-3, ETA), docosapentaenoic acid (C22:5 n-3, DPA), tetracosapentaenoic acid (C24:5 n-3), and tetracosahexaenoic acid (C24:6 n-3).

response. While ARA is a precursor of pro-inflammatory, immunosuppressive, and pro-thrombotic eicosanoids, EPA competes with ARA for lipoxygenase (LOX) and cyclooxygenase (COX) enzymes to generate functionally less intense and anti-thrombotic mediators [10]. Furthermore, EPA and DHA are precursors of resolvins and DHA is a precursor of protectins and maresins. These lipid mediators are collectively called as specialized pro-resolving mediators and have a relevant role in the inflammation resolution and homeostasis restoring [15]. In conjunction, these observations traduce an anti-inflammatory and pro-resolving potential of EPA and DHA (**Figure 2**).

Moreover, EPA, DHA, and their metabolites can exert anti-inflammatory and metabolic effects by modulating the activity of transcriptional factors, such as nuclear kappa B factor (NFκB), nuclear factor E2-related factor 2 (Nfr2), peroxisome proliferator-activated receptor (PPAR), and sterol regulatory element-binding proteins (SREBP). Due to their abilities, EPA and DHA can influence the transcription of genes enrolled in inflammation, cell survival, oxidative stress, and in carbohydrate and lipid metabolism [16]. Some of the EPA and DHA functions arise from the capacity of these n-3 PUFAs (mainly DHA) to interfere in protein receptors signaling by disrupting lipid rafts, membrane microdomains rich in saturated FAs (mainly cholesterol) who confer a rigidity needed for some protein dimerization through the fluid cell membrane [17, 18].

Due to biological properties, the importance of EPA and DHA for human health has been highly discussed and investigated by basic, translational, and epidemiologic scientists. However, studies on lipids and their biological relevance are not limited to n-3 PUFAs or other individual lipids, but also include the analysis of all

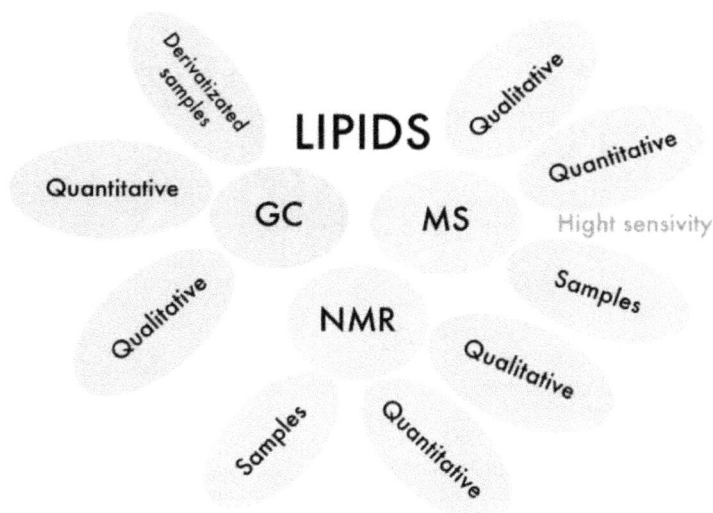

Figure 3.
The most common analytical techniques used in analyses of lipids and lipidomes are gas chromatography (GC), mass spectrometry (MS), and nuclear magnetic resonance (NMR) spectroscopy. These techniques show some vantages and weaknesses and could be used in combination with other techniques in so-called hyphenated bioanalytical methods. All enable qualitative and quantitative analyses of lipids, but GC needs additional step in sample preparation as to increase the volatility of the compounds; thus, not all lipids could be analyzed by GC. Also, GC requires greater sample quantities when compared to MS, which is most sensitive. MS analyses require the use of ionization techniques, such as electron and chemical ones for gas samples, while electrospray ionization (ESI) and matrix-assisted laser desorption/ionization (MALDI) are usually applied for liquid and solid samples. NMR is the only nondestructive technique and allows the noninvasive lipid analysis in intact cells and tissues, and enables to investigate changes in lipid and dynamic structures in biochemical cell functionalization, but it is not sufficiently sensitive and universal when compared to MS.

lipid species from a biological sample—the lipidome. Because lipids are intermediates and even signaling molecules of metabolic pathways, the lipidomic response (change of the lipidome pattern of a biological sample) to nutritional, pharmacological, or any intervention (i.e., surgery, exercise) treatments can reflect their biological effects [5]. Studies on lipidome can also add to the knowledge on the lipid content of a nutritional source (i.e., fish) aiming to found ones with the high n-3 PUFAs, for instance. These are examples of many applications of lipid analysis in biological systems.

There are several tools available for the study of individual lipids and lipidome (the total lipid content in a cell or an organism), all with their advantages and limitations (**Figure 3**). Understanding these points is essential for the application of that most appropriate techniques to answer a scientific question on lipids within a biological model. Often little familiarity with these analytical resources may limit the validity of the results. This chapter aims to present and discuss different tools available for different applications in the study of lipids aiming to assess biological hypothesis, with focus on nutrition and metabolism aspects.

2. Gas chromatography: principles, strengths, and weaknesses

According to the principles of chromatographic techniques, the gas chromatography (GC) is applied when aimed to separate organic compounds from a mixture in the gas form. For this purpose, the GC uses interaction among the sample components and the stationary phase and the mobile gas phase. After lipid extraction, the samples (lipid mixture) are usually liquids and must be exposed to a high temperature at the gas chromatograph entrance (injector). Vaporized, the samples are carried by a gas, which is usually a nonheavy and inert gas (i.e., hydrogen, helium), through a long capillary column containing a high or low polarity material (stationary phase) [19].

The gaseous compounds generated from the vaporized sample interact with the stationary phase what allows each compound to elute/separate at a different time (retention time). Because GC considers both chemical and physical properties of the vaporized compounds, those with more chemical affinity to the stationary phase will take longer time to be removed from the column and the temperature will influence the overall process. This explains why the column stays in an oven, which is programmed to work at different temperature ranges (i.e., temperature programming) in which the compounds are carried out by the gas according to their boiling point until they get to an electronic detector [20].

At the end of GC analysis, the electronic detector generates a chromatogram based on retention time by intensity. This allows a qualitative identification of the lipid compounds by comparing their retention times with certified standard using the flame ionization detector (FID) or by deduction of spectra information using a mass spectrometer as detector. Lipid quantification can also be performed using analytical procedures of external or internal certified standard in GC analysis [21].

Main points to be considered when assessing FAs by GC analysis are the carrier gas flow rate, column length, and the temperature because these can influence the order or retention time of the lipid compounds and then must be precisely standardized [22]. The column length of the stationary phase influences the resolution of the analytes, once the number of theoretical plates (hypothetical zone in which two phases establish an equilibrium with each other) is respectively high in longer column. As fat and oils have high boiling points not supported by the stationary phase, a previous derivatization reaction step is required after lipid extraction from the biological sample, in which triacylglycerol and free fatty acids are transformed into their respective free fatty esters with lower boiling points (transesterification/esterification reaction) [23]. Several methods are available for FAs derivatization [24], and the most applied ones are described in the 969.33 AOAC's method [25].

Particularly for cholesterol analysis, the samples preparation must consider a derivatization reaction. This allows to block protic sites of steroids obtained after an unsaponifiable lipid extraction [26] had been performed, and also, to diminish dipole-dipole interactions, to increase the volatility of the compounds, and to generate products with reduced polarity. Cholesterol derivatization is usually achieved by using trimethylsilyl compounds as reagents (silylation reaction). A common method for this purpose is described by Bowden and collaborators [27], in which *N,O*-bis(trimethylsilyl-trifluoroacetamide/trimethylchlorosilane)—BSTFA/TMCS is used.

Nowadays, other more modern analytical tools than GC (next-generation techniques) do not require sample derivatization for lipid analysis. Needed lipid derivatization can be then consider a quite limitation step of the technique. In comparison with next-generation techniques, GC also implies in using greater sample quantities. This may be the main limitation in biological assays, which usually lead with restricted sample amounts. Nevertheless, by using certified standard and a powerful detector as FID, GC has the advantage to allow a precise and complete (by burning every compound, no one is lost in the detection) quantification of lipid compounds from biological samples, not always achieved by the other analytical techniques. In this context, GC continues to be accepted as an efficient and simple technique for FA and sterol analyses, mainly when combined with mass spectrometry (MS, detailed later in this chapter).

2.1 Gas chromatography: application in biological assays

In biological issues, GC is largely applied to assess the FA and cholesterol contents in animal models or human fluids and tissues, as biological markers of FA ingestion and cell incorporation. The technique is a powerful tool in studies assessing the effect of FA supplementation on a specific biologic response. For instance, the endogenous synthesis of EPA and DHA from ALA is low in humans, who have in the ingestion, oily fishes as the most relevant source. Therefore, studies on n-3 PUFAs have been focused on the effect of fish ingestion or fish oil/EPA/DHA supplementation in several clinical conditions, and cell and disease models. In such studies, the treatment compliance or effectiveness can be reflected by the cell or circulate contends of n-3 PUFAs [28]. Furthermore, GC can be applied to validate data generated by other lipidomic techniques.

A practical example in using GC for treatment compliance is the study of Nogueira et al. [29] assessing the effect of n-3 PUFAs supplementation in patients with nonalcoholic steatohepatitis against placebo (mineral oil). In this study, GC highlighted a similar increase in n-3 PUFAs plasma in both n-3 PUFAs- and placebo-treated patients. Because the authors have controlled compliance of n-3 PUFAs, they were able to discover off-protocol intake of PUFAs by some patients from the placebo group. When studying biochemical markers of lipid intake and cell incorporation, the biological sample nature matters. For instance, plasma and red blood samples can reflect periods of weeks and months of FAs ingestion and their effects, respectively, while the adipose tissue is the reference method, once it reflects such variables for years [28].

The study of Ravacci et al. [30] can illustrate the use of GC to assess treatment compliance. Applying this technique, the authors were able to demonstrate that the treatment of a lineage of breast cancer cell overexpressing HER-2 with DHA increased its availability in the cell membrane and was associated with the disruption of surface lipid rafts that sustained cell signal for survival. Regarding the use of GC for data validation, a practical example is the study of Ouldamer et al. [31], which applied the technique to validate the fatty acid information generated by the ^{1}H NMR analysis on the PUFAs n-3 DHA and EPA content in the adipose tissue of mammary tumor model in rats exposed to controlled dietary intake of lipids.

3. Mass spectrometry: principles, strengths, and weaknesses

Modernization of MS used for lipid analysis raised the concept of lipidomics. Lipidomics is an emerging science that aims to analyze the total lipid content found in a cell or tissue (lipidome) through the application of analytical chemistry principles and techniques. As a part of the omics sciences, the processes applied in the lipidomic analysis are analogous to those applied in other life-building macromolecules, such as deoxyribonucleic acid (DNA), ribonucleic acid (RNA), and proteins, called as genomics, transcriptomics, and proteomics, respectively [32].

The basic principle of the MS technique is founded on the detection of the abundance of ions by their mass/charge ratio (m/z). To allow the analysis, such ions of compounds are generated by suitable methods and ions are separated according to their m/z. Ionization techniques can break some sample's molecules into charged fragments and are chosen according to the physical state of the sample. Also, the efficiency of various ionization mechanisms for the unknown species might help when picking the most appropriate ionization technique. The most common ones for gaseous samples are the electron and chemical ionizations, while for liquid and solid samples, the electrospray ionization (ESI) and matrix-assisted laser desorption/ionization (MALDI) are usually applied [33].

Advances in ESI-MS and MALDI-MS have greatly facilitated lipidomic analysis [34] and enabled a great progress in lipid metabolic discoveries. This is because ESI is one of the softest ionization techniques, in which some complex dimers and solvent adducts can also be detected at the end. The efficiency of lipid ionization in ESI is incomparably higher than achieved by other traditional MS ion sources. MALDI-MS counts on a good solubility of analytes (lipids) and a matrix (for example, 2,5-dihydrobenzoic acid) in organic solvents, and provides excellent signal-to-noise ratio and reproducibility [32].

Mass spectrometers are made from three components: the ion source (1), which converts a sample into ions that are targeted through the mass analyzer (2) and run into the detector (3). The mass analyzer acts as ions organizer (classifier) using ion m/z ratios. This component accelerates ions as they face a strong electromagnetic field. The detector measures charged particles, such as an electron multiplier [35], and the abundances of each ion present in a sample are reported.

An advantage of MS is its high sensibility. A detection limit, expressed in concentration units, goes from a mol L^{-1} to as low as fmol L^{-1} and surely shall improve as the instruments modernize. For example, the instrument response factor for any individual molecular species detected is essentially identical within experimental error after ^{13}C deisotoping if the analysis is performed properly [34]. Also, MS ion source can act as a separation device if set to selectively ionize just a certain lipid class. Thus, it is feasible to analyze different classes of lipids and individual molecular species with high efficiency without prior chromatographic separation. Nevertheless, depending on the analysis aims, MS can also be combined with a chromatography system, as GC (early mentioned) and liquid chromatography (LC) [35].

Data obtained by MS are displayed as spectra of the relative abundance of detected ions as a function of the corresponding m/z. By correlating the known masses (e.g., an entire molecule) to the identified masses, or through the compounds deposited characteristic fragmentation pattern, MS are used to identify compounds. The MS are also used to determine the elemental or isotopic signature of a sample, the masses of particles and molecules, and to elucidate the chemical structures of molecules [36]. Database platforms, such as LIPIDMAPS, LIPID Bank, LIPIDAT, Cyberlipids, and Lipidomics expertise platform, can help to identify the lipid molecules. Then, interpretation of MS-obtained lipid data must be conducted in accordance with the literature [7].

When assessing the entire lipidome profile, i.e., lipidomics by MS or nuclear magnetic resonance (NMR, detailed later in this chapter), big-data information is generated. Therefore, lipidomics require multistatistic tools for data interpretation. The additional information to MS lipidomics is mapping of the lipid pathway. For example, diacylglycerol is an essential precursor for glycerophospholipid and glycerolipid synthesis in eukaryotes [5].

Manual data interpretation using publicly available databases (i.e., KEGG pathways and the LipidMAPS databases) may add in to lipidomic results and provide meaningful biological context to data understanding from biological point of view. Indeed, using bioinformatics software platform, one can understand the changes in lipid composition and content, and understand adaptive or pathological changes in lipid metabolism. Lipids form networks, which are used to build their inter-relationships and connect them based on known metabolic pathways. Also, these relationships and the determined quantities of lipids are used to calculate the possible contributions to the production of a particular lipid class in the network, and the masses calculated are compared with the masses determined from the lipidomic MS data.

Several parameters involving the metabolic pathways can then be derived from computational simulation, such as those associated with enzymatic activities, as those analyzed by a lipid expertise, i.e., known principles of lipid biochemistry to calculate indexes of fatty acid unsaturation, fatty acyl chain length, or fatty acid precursor/product ratios to gain insight into the function of fatty acid remodeling or other relevant lipid metabolic pathways [5, 37]. Some useful tools that can be used for this purpose are the public platforms MetaboAnalyst (available from http://www.metaboanalyst.ca), VANTED, and MAVEN [37].

Once lipids have a high discrepancy of m/z within their categories and are susceptible to ion cleavage, the main disadvantage of MS in lipid analysis is that some compounds from a mixture may be determined as the same ion and incorrectly identified. Furthermore, lipid quantification by MS may be weakened by the loss of ion information due to the random collision of lipid molecules that may preclude that all of these get to the detector, the differing abilities of lipid species to form ions and hence varying signal intensity, and the ion-quenching phenomena. The last can occur when the signal from poor ionizing lipids is quenched by more easily ionized species (therefore suppressing the former signal), which is quite avoided by prior separation of lipid species for accurate quantitation or the use of specialized MS [38]. Altogether, these factors result in a loss of sensitivity for some nonpolar lipid metabolites.

It is worth to note that the limitations in identification and quantification of lipid species by MS described above have been minimized with advances of the technique (i.e., target MS). Currently, this analytic tool is considered accurate for characterization of lipids and the most efficient one to assess lipidomes.

3.1 Mass spectrometry: application in biological assays

In biological assays, lipidomics-MS analysis is highly applied to generate information related to metabolism and biological responses, once several known pathways from metabolic networks in eukaryotes involve lipids as metabolic intermediates (mainly sphingolipids, glycerophospholipids, glycerolipids, and nonesterified fatty acids [NEFAs]) or signaling molecules (mainly oxysterols) [5]. For instance, changes of a lipidome profile can be identified by MS, allowing the interpretation of biological responses to external interferences (i.e., by comparing the lipidome before and after a medication) or enrolled in the pathophysiology of diseases (in comparison with healthy status) [5, 32, 39].

The ionization technique applied is a relevant point to be considered when designing studies for lipid assessment in biological samples using MS. For instance, MALDI can be used to analyze changes of lipid and their metabolites in single genetically identical cells from the RAW264.7 lineage after lipopolysaccharide (LPS) stimulation, using a Fourier transform ion cyclotron resonance mass spectrometer (FTICR MS). MALDI analysis was chosen because single cells on a plate using a histology-directed workflow can increase the number of cells analyzed. Furthermore, the speed of MALDI-IMS enables high spatial resolution and high-throughput single-cell analysis. Combined with the high sensitivity of FTICR MS, hundreds of lipids can be measured from a large population of single cells (>100) in a few hours.

Tandem MS measurements (i.e., through precursor ion scanning and neutral loss scanning experiments) are useful for biological assays requiring the identification of all lipid molecular species. These methodologies are usually better than full scan MS because they apply sequential analyzers and are often associated with a target analysis (i.e., aiming to study a molecule species). This allows high sensitivity and enhanced signal/noise ratio, facilitating the characterization of minor but biologically relevant lipid species [40].

One example of the tandem analysis application is the work of Slatter et al. [41]. By using LC-MS/MS (tandem MS), they were able to characterize the lipidomic network of human platelets, where nearly 200 oxidized species were identified. These minacious data provided by the methodology allowed to display a direct link between innate immunity and mitochondrial bioenergetics in human platelets. Procedures enabling to achieve this conclusion from generated data included the selection of lipids upregulated under thrombin activation and the analysis on temporal dynamics of their generation, monitoring precursor-to-product ion transitions in multiple reaction monitoring (MRM) modes.

Also, through tandem MS, Morgan et al. [42] have proposed a novel role for 12/15-lipoxygenase in regulating autophagy. They have used LC/ESI/MS/MS in a target approach to determine the levels of 1,2-dimyristoyl-sn-glycero-3-phosphoethanolamine (DMPE), 1,2-dimyristoyl-sn-glycero-3-phosphocholine (DMPC), 1,2-dimyristoyl-sn-glycero-3-phosphate (DMPA), and 1,2-dimyristoyl-sn-glycero-3-phospho-(1'-rac-glycerol) (DMPG), using comparison technique with internal standards. In addition, the 1,2-dimyristoyl-sn-glycero-3-phosphoserine (DMPS) was determined by product ions and the analysis of cholesteryl esters was performed.

4. Nuclear magnetic resonance spectroscopy: principles, strengths, and weaknesses

Along with other analytical tools available for lipidome investigations, NMR spectroscopy allows identification of characteristic signals from the different classes of lipids and provides their successful quantification [43, 44]. The technique facilitates the analysis of hundreds of metabolites in a single sample with great advantage because there is no need for a previous sample treatment [8].

The principle of NMR spectroscopy is based on the physical resonance phenomenon in which spin-active nuclei in a strong static magnetic field respond to a perturbation (radiofrequency waves) by producing an electromagnetic signal with a characteristic frequency, which matches magnetic field observed by a given nucleus. This process of resonance happens when the oscillation frequency matches the intrinsic frequency of the nuclei, which depends on the strength of the static magnetic field, the chemical environment, and the magnetic properties of the isotope involved [45].

In a practical way, NMR spectroscopy provides information of the number (integrals) of magnetically distinct atoms (chemical shift of the resonance frequencies and peak splitting due to the coupling constants J or dipolar couplings between nuclear spins in the sample) of the studied isotope and provides all necessary information for determination of the structure of unknown molecules. Several nuclei can be studied by NMR techniques, but the most commonly available ones are hydrogen-1 and carbon-13. The most common experiments for lipid analysis by NMR are ^{1}H, ^{13}C, ^{31}P, and the bidimensional experiments involving ^{1}H-^{1}H and ^{1}H-^{13}C [45].

Usually, an NMR experiment starts with insertion of a liquid sample into the magnet, then, short radio-frequency pulses (from an electronic device named probe) are applied, and all emitted frequencies from the same type of nuclei are registered and reported as signals with a given chemical shift, multiplicity, and intensity. Also, multidimensional NMR as well as solid-state NMR has emerged to provide additional and relevant information on sample composition [45].

Also, the exact ratio of specific fatty acids in the lipid samples and their iodine values could be calculated considering integral values corresponding to characteristic peaks with the help of the corresponding spectral information and the existing references [46]. This type of experiment works as a relative quantification. Absolute and relative quantification experiments by NMR are possible; however, it is necessary to take care of some precautions. Direct quantitative information by NMR is due to the fact that the signal intensity of each resonance in the NMR spectrum is directly proportional to the number of spins associated with the particular resonance [38]. Thus, no standard with chemical similarity to the studied compounds is required as in other analytical methods; however, one certified standard must be used. This can be performed through relative quantification using ERETIC. For absolute quantification also, a certified standard is required now as an internal standard in a known concentration. For both methods, the pulse sequence needs to be calibrated to 90° to be sure that the spectral response is completely real, and it means that the longitudinal relaxation time (T_1) of spins is entirely returned [38]. Typically, this is achieved by waiting five times the longest T_1 (at five times T_1 approximately 99.3% of the equilibrium value is re-established) between two scans.

Proton magnetic resonance spectroscopic imaging (^{1}H-MRSI) has a major role in lipid assays, mainly used in the medical area with extreme importance for *in vivo* sampling. Both profiling and ratio quantifications are possible by the obtained spatial resolved spectra. The presence of so many compounds in living biological samples may require water or other signal suppression experiments to be performed in order to obtain better resolution on the target metabolites. The same approach is used in NMR samples but with greater implications due to lack of sample pretreatment [47].

Compared to the MS method, NMR technique is less sensitive and limited by the overlapping of signals in either, ^{1}H NMR or ^{31}P NMR, and also by the low natural abundance of ^{13}C for ^{13}C NMR. On the other hand, NMR is a nondestructive sample technique that allows a high analytical reproducibility, an easy identification of molecular moieties, and with relatively easy to get information on molecular dynamics [8, 38]. Furthermore, NMR does not require a standard curve or molecule species for quantitative measuring. Therefore, this technique has been emerging as a promising approach for more accurate and faster quantitative analysis of lipids than other analytical methods [38]. Also, the sensitivity improvement of cryogenic probe in an equipment of 800 MHz LC-NMR is very promising in analysis of a trace amount of lipids in a faster experiment, once it is able to acquire ^{1}H NMR spectrum of approximately 1 μg sample within 30 min, whereas the current 500 MHz NMR needs 20 h or longer [38].

4.1 Nuclear magnetic resonance spectroscopy: application in biological assays

A wide variety of NMR experiments (e.g., HSQC, HMBC, TOCSY, etc.) besides the classics ^{1}H, ^{13}C, and ^{31}P NMR are being used to solve a variety of biological issues where biofluid samples such as serum, plasma, urine, cerebrospinal fluid (CSF), etc., are being investigated. More commonly used are ^{1}H, ^{13}C, and ^{31}P NMR experiments, which bring rich information on lipid profiling, for example, molecular identification of fatty acid chains and phospholipid structures. Furthermore, heteronuclear and multidimensional experiments can be used to elucidate lipid profiling information by signal interpretation and also using comparisons with databases. The ^{13}C NMR is also a complementary tool that can be used for fatty acyl residue identification [38].

Once NMR allows the noninvasive lipid analysis in intact cells and tissues, the technique prevents losses of chemical information in the analyte environment. This fact, together with the high sensibility of NMR to molecular dynamics (in timescales from picoseconds to seconds), enables to investigate changes in lipid and dynamic structures in biochemical cell functionalization. The experiment used for this application is the diffusion ordered spectroscopy (DOSY), which enables to separate signals according to their diffusion coefficients and then add chromatography-like capabilities to NMR [38, 48].

Lipoproteins consist mainly from cholesterol esters and triacylglycerols surrounded by a hydrophilic layer, which comprehend phospholipids, cholesterol, and proteins [8]. Lipoproteins perform the lipid transportation in blood circulation through the exogenous (dietary lipids) and the endogenous (liver-synthetized lipids) channels. The endogenous transportation begins in the liver through the production of a very low-density lipoprotein (VLDL). After being secreted into the bloodstream, VLDL interacts with other lipoproteins, through collisions, in which the contact with the high-density lipoprotein (HDL) is highlighted.

Kostara et al. [49] have found how blood lipoproteins influence the progression of coronary heart disease (CHD) by comparing the lipid profiles of atherogenic (non-HDL) and atheroprotective (HDL) lipoproteins from patients with CHD with those from patients with normal coronary arteries (NCA). They analyzed the lipid extracts of these lipoproteins using ^{1}H NMR experiments and statistical analysis and identified the potential target-lipid biomarkers for the early evaluation of the CHD onset. Furthermore, Lopes et al. [50] were able to find that circulating HDL increases, and LDL and VLDL decrease in obese patients after bariatric surgery by using DOSY experiments to monitor these lipoproteins. Notably, lipoprotein investigations and quantitative analysis of lipids can be performed using NMR of the same sample [51].

Also, selective recoupling of dipolar and chemical-shift interactions removed by magic-angle spinning NMR in the solid state allows the characterization of regulatory interactions, dynamics, and ion channels within biological membranes [52].

In this scenario, the NMR application has contributed to obtaining of important data on the structure and turnover of lipid species and the composition of lipids in cells, and to characterize pathways enrolled in lipid synthesis/transport and degradation [53, 54]. Also, the high-resolution magic-angle spinning NMR (HR-MAS NMR) has been applied to global lipidomic studies [52].

Besides the identification of lipid species and dynamics, NMR can be used for reliable quantification of lipid mixtures obtained from tissues, body fluids, and cell cultures [40, 55]. It can be allied to the bioinformatic tools available to a better quantitative analysis of lipid profiles [56]. For instance, using ^{1}H NMR and ^{31}P NMR, Fernando et al. [57] were able to identify an over-accumulation of lipids associated with the pathophysiology of ethanol-induced liver steatosis accompanied by mild inflammation.

Also, quantification can be used in magnetic resonance imaging (MRI) experiments as Vafaeyan et al. [47] have shown. They have used a time-domain quantification method namely as subtract-QUEST-MRSI algorithm to quantify alterations of the biomarkers, i.e., lipids and other metabolism molecules species such as choline, creatine, *N*-acetyl aspartate, lactate, myo-inositol, and glutamine in multiple sclerosis subjects in comparison with control group. This research aimed to know how lesion biomarker ratios in multiple sclerosis have affected human brains, through the imaging of different brain areas, which could present lesions.

Other MRI works have found that on brain imaging, lipids tend to be an almost undesired artifact, and consequently, scientists may use the approach of selective signal suppression pulses such as adiabatic frequency selective, spatial-spectral lipid suppression, or broadband outer volume suppression bands [58]. Trauner et al. [59] have used a dynamic saturation transfer technique in MRI experiments to assess dynamic Pi-to-ATP exchange parameters in nonalcoholic fatty liver disease (NAFLD) and steatohepatitis (NASH) aiming to report alterations of hepatic lipid, cell membrane, and energy.

5. Final considerations

Lipids *per se* exert several relevant biological functions making the single knowledge of the lipidome profile from a biological sample highly informative by itself. For instance, sphingolipids and glycerophospholipids are important components of the cell membrane and then can affect several cellular functions. Disorders of sphingolipid metabolism are associated with lysosomal storage diseases and of lysoglycerophospholipid by phospholipase A2 activation are associated with lipotoxicity and inflammation. Accumulation of triacylglycerol (a glycerophospholipid) is associated with lipotoxicity and insulin resistance, and the NEFA profile is a useful indicator of lipid metabolism and can add to understanding on molecular mechanisms underlying the metabolic syndrome [5].

Therefore, lipidomic tools are particularly useful to identify and understand changes in metabolic pathways and the underlying mechanisms enrolled in the pathophysiology of human health, such as metabolic diseases. One practical example is data from Meikle et al. [60] study that measured 259 lipid species in plasma samples from prediabetic, diabetic and normal glucose tolerant patients, including sphingolipids, phospholipids, glycerolipids, and cholesterol ester. The authors used electrospray ionization-tandem mass spectrometry in previous precursor ion and neutral loss scans on control plasma extracts, MRM experiments for the major species of each lipid class identified in plasma, and quantification using internal standards. These approaches highlighted that metabolic pathways altered in type 2 diabetes include a deregulation of lipid homeostasis, characterized by abnormal plasma-free fatty acids accumulation.

In lipidomic studies, beyond the care of equipment calibration and accuracy of the experiments, special cares of analytical procedures must be planned to have accurate information of data. The statistical recourses are necessary to process the data, but, also lipid knowledge is required for correct interpretation in all cases. The choice of the most suitable lipidomic tool to be used for a specific biological assay is closely linked to the study aim. Next-generation techniques (MS and NMR) can provide detailed lipid information to assess more elaborated scientific questions. However, thousands of individual lipid molecular species are present in cells implying that no single technique can effectively study all the lipid species [38]. When possible, combining techniques can be the best choice, because one can compensate for the limitation of the other, and bring complementary information and/or can validate the previous analysis data.

Usually, combined lipidomic techniques are applied for data validation. For instance, data obtained by shotgun lipidomics (direct infusion MS) can be validated using LC-MS-based analyses and *vice versa*. Other methods, including NMR, or chromatography-based analysis might be used to validate the total lipid content of a lipid class [5]. However, the combined use of lipidomic techniques can also be useful to improve the data information on the lipids from biological samples. For instance, to assess lipid changes during the response of hypoxia stress to a treatment in cervical cancer-derived cells (HeLa cells), Yu et al. [40] applied NMR technique for the phospholipid profile analysis and MS for phospholipids characterization. Also, Whiley et al. [61] investigated the plasma phosphatidylcholine metabolism using NMR and MS to obtain a fingerprint of three phosphatidylcholines (PC) molecules that significantly decrease in individuals with Alzheimer's disease compared to healthy controls. Then, LC-MS and NMR were used for phosphatidylcholine and fatty acyl side chain validation and for total plasma choline validation, respectively. The study of Whiley et al. [61] illustrates the scientific value in combining different lipidomic tools to obtain complementary information and reinforce validation of the obtained data.

In conclusion, all available tools for lipidomic studies in biological samples have several advantages and limitations that can be overcome when combining more than one technique. Because this practice involves the availability of complex technologies and skilled labor, it is not always possible. In this scenario, the use of mass spectrometry alone can be the best alternative currently available when technique combination is impossible. However, NMR has a high potential and, in the future, may be expected to answer issues that MS is quite limited to do.

Acknowledgements

This research counted on grant received from the Brazilian agency—*Fundação de Amparo à Pesquisa do Estado de São Paulo*—FAPESP (Sao Paulo Research Foundation, grant number 2018/06510-4).

Conflict of interest

Authors declare no conflict of interests.

Author details

Banny Silva Barbosa Correia[1], Raquel Susana Torrinhas[2], William Yutaka Ohashi[3,4] and Ljubica Tasic[3*]

1 Chemistry Institute, University of Sao Paulo (USP), Sao Carlos, Sao Paulo, Brazil

2 Department of Gastroenterology, Surgical Division (LIM 35), University of Sao Paulo (USP), School of Medicine, Sao Paulo, Brazil

3 Chemistry Institute, Campinas State University (UNICAMP), Campinas, Sao Paulo, Brazil

4 Agilent Technologies Brasil Ltda., Barueri, Sao Paulo, Brazil

*Address all correspondence to: ljubica@iqm.unicamp.br

IntechOpen

References

[1] Christie W. What is a Lipid?
[Internet]. 2013. Available from: http://
lipidlibrary.aocs.org/Primer/content.
cfm?ItemNumber=39371&navItemNum
ber=19200 [Accessed: 19 May 2017]

[2] Dewich PM. Medicional Natural
Products—A Biosynthetic Approach.
2nd ed. Chichester: John Wiley & Sons;
2002. p. 507. ISBNs: 0471496405

[3] Fahy E, Subramanian S, Brown HA,
Glass CK, Merrill AH, Murphy RC, et al.
A comprehensive classification system
for lipids. Journal of Lipid Research.
2005;**46**(5):839-862. DOI: 10.1194/jlr.
E400004-JLR200

[4] Quehenberger O, Armando AM,
Brown AH, Milne SB, Myers DS,
Merril AH, et al. Lipidomics reveals a
remarkable diversity of lipids in human
plasma. Journal of Lipid Research.
2010;**51**(11):3299-3305. DOI: 10.1194/jlr.
M009449

[5] Han X. Lipidomics for studying
metabolism. Nature Reviews
Endocrinology. 2016;**12**(11):668-679.
DOI: 10.1038/nrendo.2016.98

[6] LIPID MAPS. Lipid Classification
System [Internet]. 2017. Available
from: http://www.lipidmaps.org/data/
classification/LM_classification_exp.
php [Accessed: 19 May 2017]

[7] Brugger B. Analysis of the
lipid composition of cells and
subcellular organelles by eletrospray
ionization mass spectrometry.
Annual Review of Biochemistry.
2014;**83**:79-98. DOI: 10.1146/
annurev-biochem-060713-035324

[8] Rolim AEH, Henrique-Araújo R,
Ferraz EG, Dultra FKAA, Fernandez LG.
Lipidomics in the study of lipid
metabolism: Current perspectives
in theomic sciences. Gene.
2015;**554**(2):131-139. DOI: 10.1016/j.
gene.2014.10.039

[9] Calder PC, Deckelbaum RJ. Dietary
lipids: More than just a source of
calories. Current Opinion in Clinical
Nutrition and Metabolic Care.
1999;**2**(2):105-107

[10] Calder PC. Marine omega-3 fatty
acids and inflammatory processes:
Effects, mechanisms and clinical
relevance. Biochimica et Biophysica
Acta. 2015;**1851**(4):469-484. DOI:
10.1016/j.bbalip.2014.08.010

[11] Muro E, Atilla-Gokcumen GE,
Eggert US. Lipids in cell biology:
How can we understand them
better? Molecular Biology of the Cell.
2014;**25**(12):1819-1823. DOI: 10.1091/
mbc.E13-09-0516

[12] Parton DL, Klingelhoefer JW,
Sansom MS. Aggregation of model
membrane proteins, modulated by
hydrophobic mismatch, membrane
curvature, and protein class. Biophysical
Journal. 2011;**101**(3):691-699. DOI:
10.1016/j.bpj.2011.06.048

[13] Simons K, Toomre D. Lipids rafts
and signal transduction. Nature.
2000;**1**(1):31-39. DOI: 10.1038/35036052

[14] Waitzberg DL, Torrinhas
RS. Fish oil lipid emulsions and
immune response? What clinicians
need to know. Nutrition in Clinical
Practice. 2009;**24**(4):487-499. DOI:
10.1177/0884533609339071

[15] Bannenberg GL, Chiang N, Ariel
A, Arita M, Tjonahen E, Gotlinger KH,
et al. Molecular circuits of resolution:
Formation and actions of resolvins and
protectins. The Journal of Immunology.
2005;**174**(7):4345-4355. DOI: 10.4049/
jimmunol.174.7.4345

[16] Calder PC. Functional roles of
fatty acids and their effects on human
health. Journal of Parenteral and Enteral
Nutrition. 2015;**39**(1 Suppl):18S-32S.
DOI: 10.1177/0148607115595980

[17] Shaikh SR, Rockett BD, Salameh M, Carraway K. Docosahexaenoic acid modifies the clustering and size of lipid rafts and the lateral organization and surface expression of MHC class I of EL4 cells. The Journal of Nutrition. 2009;**139**(9):1632-1639. DOI: 10.3945/jn.109.108720

[18] Siddiqui RA, Harvey KA, Zaloga GP, Stillwell W. Modulation of lipid rafts by n-3 fatty acids in inflammation and cancer: Implications for use of lipids during nutrition support. Nutrition in Clinical Practice. 2007;**22**(1):74-88

[19] Bertsch W. Two-dimensional gas chromatography. Concepts, instrumentation, and applications–part 2: Comprehensive two-dimensional gas chromatography. Journal of Separation Science. 2000;**23**(3):167-181. DOI: 10.1002/(SICI)1521-4168(20000301)23:3<167::AID-JHRC167>3.0.CO;2-2

[20] HM MN, Miller JM. Basic Gas Chromatography. 2nd ed. Chichester: John Wiley & Sons; 2009. p. 256. DOI: 10.1002/9780470480106

[21] Saint Laumer J-Y, Cicchetti E, Merle P, Egger J, Chaintreau A. Quantification in gas chromatography: Prediction of flame ionization detector response factors from combustion enthalpies and molecular structures. Journal of Analytical Chemistry. 2010;**82**(15):6457-6462. DOI: 10.1021/ac1006574

[22] Seppanen-Laakso T, Hiltunen R. Analysis of fatty acids by gas chromatography, and its relevance to research on health and nutrition. Analytica Chimica Acta. 2002;**465**(1-2):39-62. DOI: 10.1016/S0003-2670(02)00397-5

[23] Nikelly JG. Gas chromatography of free fatty acids. Journal of Analytical Chemistry. 1964;**36**(12):2244-2248. DOI: 10.1021/ac60218a007

[24] Ostermann AI, Müller M, Willenberg I, Schebb NH. Determining the fatty acid composition in plasma and tissues as fatty acid methyl esters using gas chromatography–a comparison of different derivatization and extraction procedures. Prostaglandins, Leukotrienes and Essential Fatty Acids. 2014;**91**(6):235-241. DOI: 10.1016/j.plefa.2014.10.002

[25] Horwitz W. Official Methods of Analysis of AOAC International. 17th ed. Gaithersburg, Maryland: AOAC International; 2000

[26] Azevedo-Meleiro CH, Rodriguez-Amaya DB. Confirmation of the identity of the carotenoids of tropical fruits by HPLC-DAD and HPLC-MS. Journal of Food Composition and Analysis. 2004;**17**:385-396. DOI: 10.1016/j.jfca.2004.02.004

[27] Bowden JA, Colosi DM, Mora-Montero DC, Garrett TJ, Yost RA. Enhancement of chemical derivatization of steroids by gas chromatography/mass spectrometry (GC/MS). Journal of Chromatography B. 2009;**877**(27):3237-3242. DOI: 10.1016/j.jchromb.2009.08.005

[28] Silva V, Barazzoni R, Singer P. Biomarkers of fish oil omega-3 polyunsaturated fatty acids intake in humans. Nutrition in Clinical Practice. 2014;**29**(1):63-71. DOI: 10.1177/0884533613516144

[29] Nogueira MA, Oliveira CP, Ferreira Alves VA, Stefano JT, Rodrigues LS, Torrinhas RS, et al. Omega-3 polyunsaturated fatty acids in treating non-alcoholic steatohepatitis: A randomized, double-blind, placebo-controlled trial. Clinical Nutrition. 2016;**35**(3):578-586. DOI: 10.1016/j.clnu.2015.05.001

[30] Ravacci GR, Brentani MM, Tortelli TJ, Torrinhas RS, Saldanha T, Torres EA, et al. Lipid raft disruption by

docosahexaenoic acid induces apoptosis in transformed human mammary luminal epithelial cells harboring HER-2 overexpression. The Journal of Nutrition Biochemistry. 2013;**24**(3):505-515

[31] Ouldamer L, Nadal-Desbarats L, Chevalier S, Body G, Goupille C, Bougnoux P. NMR-based lipidomic approach to evaluate controlled dietary intake of lipids in adipose tissue of a rat mammary tumor model. Journal of Proteome Research. 2016;**15**(3):868-878. DOI: 10.1021/acs.jproteome.5b00788

[32] Wenk M. The emerging field of lipidomics. Nature Reviews Drug Discovery. 2005;**4**(7):594-610. DOI: 10.1038/nrd1776

[33] Han X, Yang K, Gross RW. Multi-dimensional mass spectrometry-based shotgun lipidomics and novel strategies for lipidomic analyses. Mass Spectrometry Reviews. 2012;**31**(1):134-178. DOI: 10.1002/mas.20342

[34] Wang C, Wang M, Han X. Applications of mass spectrometry for cellular lipid analysis. Molecular BioSystems. 2015;**11**(3):698-713. DOI: 10.1039/c4mb00586d

[35] Hoffman E, Stroobant V. Mass Spectrometry: Principles and Applications. 3rd ed. Chichester: John Whiley & Sons; 2007. p. 489. ISBN: 978-0-470-03310-4

[36] Milman BL. General principles of identification by mass spectrometry. Trends in Analytical Chemistry. 2015;**69**:24-33. DOI: 10.1016/j.trac.2014.12.009

[37] Lydic TA, Goo YH. Lipidomics unveils the complexity of the lipidome in metabolic diseases. Clinical and Translational Medicine. 2018;7(1):4. DOI: 10.1186/s40169-018-0182-9

[38] Li J, Vosegaard T, Guo Z. Applications of nuclear magnetic resonance in lipid analyses: An emerging powerful tool for lipidomics studies. Progress in Lipid Research. 2017;**68**:37-56. DOI: 10.1016/j. plipres.2017.09.003

[39] Holcapek M, Liebisch G, Ekroos K. Lipidomic Analysis. Analytical Chemistry. 2018;**90**(7):4249-4257. DOI: 10.1021/acs.analchem.7b05395

[40] Yu Y, Vidalino L, Anesi A, Macchi P, Guella G. A lipidomics investigation of the induced hypoxia stress on HeLa cells by using MS and NMR techniques. Molecular BioSystems. 2014;**10**(4):878-890. DOI: 10.1039/c3mb70540d

[41] Slatter DA, Aldrovandi M, O'Connor A, Allen SM, Brasher C, Murphy RC, et al. Mapping the human platelet lipidome reveals cytosolic phospholipase A2 as a regulator of mitochondrial bioenergetics during activation. Cell Metabolism. 2016;**23**(5):930-944. DOI: 10.1016/j. cmet.2016.04.001

[42] Morgan AH, Hammond VJ, Sakon-Nakatogawa M, Ohsumi Y, Thomas CP, Blanchet F, et al. A novel role for 12/15-lipoxygenase in regulating autophagy. Redox Biology. 2015;**4**:40-47. DOI: 10.1016/j.redox.2014.11.005

[43] Lutz NW, Sweedler JV, Wevers RA. Methodologies for Metabolomics: Experimental Strategies and Techniques. New York: Cambridge University Press; 2013. ISBN: 978-0521765909

[44] Vidal NP, Manzanos MJ, Goicoechea E, Guillén MD. Quality of farmed and wild sea bass lipids studied by 1H NMR: Usefulness of this technique for differentiation on a qualitative and a quantitative. Food Chemistry. 2012;**135**(3):1583-1591. DOI: 10.1016/j. foodchem.2012.06.002

[45] Levitt MH. Spin Dynamics: Basics of Nuclear Magnetic Resonance. 2nd ed. Chichester: John Wiley & Sons; 2008

[46] Zhang Y, Zhao Y, Shen G, Zhong S, Feng J. NMR spectroscopy in conjugation with multivariate statistical analysis for distinguishing plant origin of edible oils. Journal of Food Composition and Analysis. 2018;**69**:140-148. DOI: 10.1016/j.jfca.2018.03.006

[47] Vafaeyan H, Ebrahimzadeh SA, Rahimian N, Alavijeh SK, Madadi A, Faeghi F, et al. Quantification of diagnostic biomarkers to detect multiple sclerosis lesions employing 1H-MRSI at 3T. Australasian Physical & Engineering Sciences in Medicine. 2015;**38**(4):611-618. DOI: 10.1007/s13246-015-0390-1

[48] Dyrby M, Petersen M, Whittaker AK, Lambert L, Nørgaard L, Bro R, et al. Analysis of lipoproteins using 2D diffusion-edited NMR spectroscopy and multi-way chemometrics. Analytica Chimica Acta. 2005;**531**(2):209-216. DOI: 10.1016/j.aca.2004.10.052

[49] Kostara CE, Papathanasiou A, Psychogios N, Cung MT, Elisaf MS, Goudevenos J, et al. NMR-based lipidomic analysis of blood lipoproteins differentiates the progression of coronary heart disease. Journal of Proteome Research. 2014;**13**(5):2585-2598. DOI: 10.1021/pr500061n

[50] TIB L, Geloneze B, Pareja JC, Calixto AR, Ferreira MMC, Marsaioli AJ. Omics prospective monitoring of bariatric surgery: Roux-En-Y gastric bypass outcomes using mixed-resolved 1H NMR-based metabolomics. Journal of Integrative Biology. 2016;**20**(7):415-423. DOI: 10.1089/omi.2016.0061

[51] Barbosa BS, Martins LG, TBBC C, Cruz G, Tasic L. Qualitative and quantitative NMR approaches in blood serum lipidomics. In: Guest P, editor. Investigation of Early Nutrition effects on Long-Term Health—Methods in Molecular Biology. New York: Humana Press; 2018. pp. 365-379. DOI: 10.1007/978-1-4939-7614-0_25

[52] Gross RW, Han X. Lipidomics at the interface of structure and function in systems biology. Cell. 2011;**18**(3):284-291. DOI: 10.1016/j.chembiol.2011.01.014

[53] Sethi S, Hayashi M, Sussulini A, Tasic L, Brietzke E. Analytical approaches for lipidomics and its potential applications in neuropsychiatric disorders. The World Journal of Biological Psychiatry. 2017;**18**(7):506-520. DOI: 10.3109/15622975.2015.1117656

[54] Sethi S, Hayashi MAF, Barbosa BS, Pontes JGM, Tasic L, Brietzke E. Lipidomics, biomarkers, and schizophrenia: A current perspective. In: Sussulini A, editor. Metabolomics: From Fundamentals to Clinical Applications, Advances in Experimental Medicine and Biology. Charm: Springer; 2017. pp. 265-290. DOI: 10.1007/978-3-319-47656-8_11

[55] Gallo V, Intini N, Mastrorilli P, Latronico M, Scapicchio P, Triggiani M, et al. Performance assessment in fingerprinting and multi component quantitative NMR analyses. Analytical Chemistry. 2015;**87**(13):6709-6717. DOI: 10.1021/acs.analchem.5b00919

[56] Barrilero R, Gil M, Amigo N, Dias CB, Wood LG, Garg ML, et al. LipSpin: A new bioinformatics tool for quantitative 1H NMR lipid profiling. Analytical Chemistry. 2018;**90**(3):2031-2040. DOI: 10.1021/acs.analchem.7b04148

[57] Fernando H, Bhopale KK, Kondraganti S, Kaphalia BS, Ansari GAS. Lipidomic changes in rat liver after long-term exposure to ethanol. Toxicology and Applied Pharmacology. 2011;**255**(2):127-137. DOI: 10.1016/j.taap.2011.05.022

[58] Henning A, Schar M, Schulfe RF, Wilm B, Pruessmann KP,

Boesiger P. SELOVS: Brain MRSI localization based on highly selective T1- and B1-insensitive outer-volume suppression at 3T. Magnetic Resonance in Medicine. 2008;**59**(1):40-51. DOI: 10.1002/mrm.21374

[59] Traussnigg S, Kienbacher C, Gajdošík M, Valkovič L, Halilbasic E, Stiff J, et al. Ultra-high-field magnetic resonance spectroscopy in non-alcoholic fatty liver disease: Novel mechanistic and diagnostic insights of energy metabolism in non-alcoholic steatohepatitis and advanced fibrosis. Metabolic Liver Disease. 2017;**37**(10):1544-1553. DOI: 10.1111/liv.13451

[60] Meikle PJ, Wong G, Barlow CK, Weir JM, Greeve MA, MacIntosh GL, et al. Plasma lipid profiling shows similar associations with Prediabetes and type 2 diabetes. PLoS One. 2013;**8**(9):e-74341. DOI: 10.1371/journal.pone.0074341

[61] Whiley L, Sen A, Heaton J, Proitsi P, García-Gómez D, Leung R, et al. Evidence of altered phosphatidylcholine metabolism in Alzheimer's disease. Neurobiology of Aging. 2013;**35**(2):1-8. DOI: 10.1016/j.neurobiolaging.2013.08.001

Chapter 3

Plant Lipid Metabolism

Fatiha AID

Abstract

In plants, the synthesis of fatty acids takes place in the chloroplast and the fatty acid synthase is prokaryotic type. In plants, the structure of membrane lipids is different from that of eukaryotic cells. The membranes of the chloroplasts are essentially formed of galatolipids. This chapter will also focus on the structure and biosynthesis of fatty acids and membrane lipids in plants. Lipids of seeds are essentially composed of TAG; it would be interesting to describe their synthesis during the maturation of the seeds. Some plants contain in their reserve lipids unconventional fatty acids such as gamma linolenic acid in *Borrago officinalis* L., short-chain fatty acids C: 12 and C: 10, fatty acids with very long chains, and fatty acids that are cyclical. All of these fatty acids can have industrial and/or pharmaceutical applications.

Keywords: fatty acid, lipids, biosynthesis, plant

1. Introduction

Plants produce the majority of lipids in the world. These lipids are the main source of calories and essential fatty acids for men and animals. Plants synthesize a huge variety of fatty acids although only a few are major and common constituents [1] like palmitic, oleic, linoleic, and linolenic acids. Like other eukaryotes, lipids are necessary for the biogenesis of cell membranes, as signal molecules and especially as a source of carbon and energy. In plants, carbon, energy, and reducing power are needed for fatty acid biosynthesis derived from photosynthesis in chloroplasts [2]. This chapter will describe lipid biosynthesis in plants by signaling differences with other organisms and highlighting the specificity of plants.

Fatty acid biosynthesis in plants occurs in the chloroplasts of green tissue and in the plastids of nonphotosynthetic tissues and not in the cytosol as in the animal cell. Although *de novo* synthesis is located in the stroma, plant mitochondria are capable of limited fatty acid synthesis [3]. The plastid membranes are mainly composed of galactolipids, while those of extrachloroplast membranes consist of phospholipids as in the animal cell [4]. Fatty acids in cell membranes consist mainly of palmitic, stearic, oleic, linoleic, and linolenic acids. All double bonds are of cis type. However, in the chloroplast, phosphatidyl glycerol (PG) is acylated with an unusual acid having a trans-type double bond: $\Delta 3$ 16: 1t [5].

Photosynthetic tissues of higher plants contain 60–70% trienoic fatty acids. The so-called "C18: 3" plants are generally the most advanced families of angiosperms (pea, spinach, etc.) whose position *sn*-2 of the galactolipids is esterified exclusively by polyunsaturated fatty acids with 18 carbon atoms. The "C16: 3" plants are generally the less evolved families of angiosperms (Brassicaceae) whose position *sn*-2 of the galactolipids is esterified by polyunsaturated fatty acids with 16 or 18 carbon atoms [6].

Plant lipids have a substantial impact on the world economy and human nutrition. The majority of oils used by humans are triacylglycerols derived from seeds or fruits. Indeed, the seeds are subdivided into three categories according to their reserve. Seeds that contain more than 45% protein are called protein seeds. Starchy seeds contain more than 70% of carbohydrates like cereals. Oleaginous seeds contain more than 50% of lipids in the form of triacylglycerol esterified generally by palmitic, oleic, linoleic, and linolenic acids in majority of seeds. Some plants can produce unusual fatty acids like hydroxyl fatty acids, cyclopropane fatty acids, epoxy fatty acids, and conjugated unsaturated fatty acids in their seed oils, many of which have useful industrial applications [7]. These unusual fatty acids accumulate preferentially in triacylglycerols.

2. Fatty acid synthesis

In plants, *de novo* fatty acid biosynthesis mainly takes place in the plastidial compartment [8] from acetyl CoA, which is a direct product of photosynthesis. Plastid pyruvate dehydrogenase (EC 1.2.4.1) is the main route for a rapid and stable supply of acetyl CoA through its action on pyruvate (resulting from glycolysis or the pentose phosphate pathway). Another possible source is the import and activation of free acetate by acetyl CoA synthase (ACS, EC 6.2.1.1) [9]. The major product of FAS is palmitic acid, except the elongation of palmitic acid and the desaturation of stearic acid which take place in the chloroplast. Other changes (elongation, desaturation, hydroxylation, and epoxidation) occur mainly in the endoplasmic reticulum.

Two enzyme systems are required for fatty acid formation: acetyl CoA carboxylase (ACCase, EC 6.4.1.2) of which two forms have been identified in plants [10] and fatty acid synthase which is a multienzyme complex present in the stroma of chloroplasts [11].

2.1 Acetyl CoA carboxylase (ACCase)

The first enzyme complex is the ACCase that catalyzes an ATP-dependent carboxylation of acetyl CoA to malonyl CoA. For plants, acetyl CoA carboxylase (ACCase) directs the flow of carbon from photosynthesis to primary and secondary metabolites. Two distinct molecular forms of ACCase have been identified, a multiprotein complex and a multifunctional protein [12] (**Figure 1**).

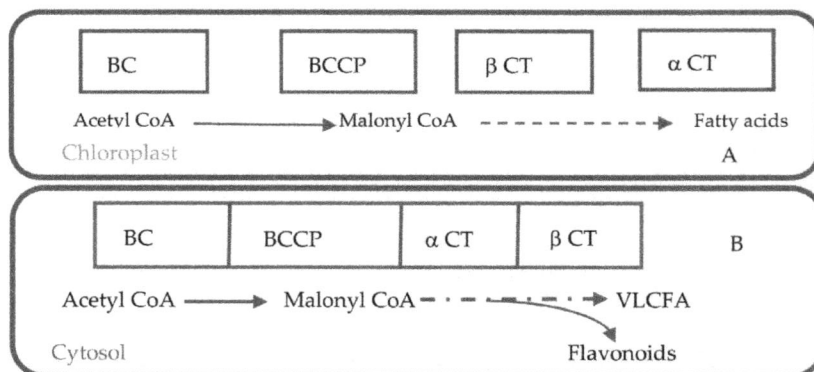

Figure 1.
Structure of the two types of ACCase. (A) The multisubunit (MS complex) ACCase and (B) the multifunctional (MF) ACCase. BCCP, biotin carboxyl carrier protein; BC, biotin carboxylase; α and β CT, α and β carboxy transferase; VLCFA, very long-chain fatty acids.

The multisubunit (MS complex) ACCase, present in plastids of all plants, except *Poaceae* and *Geraniaceae*, is involved in *de novo* fatty acid synthesis [13]. It is composed of four independent polypeptides: biotin carboxyl carrier protein (BCCP), biotin carboxylase (BC) and α and β carboxy transferase (α and β CT). The biotin carboxylase (BC) subunit catalyzes the ATP-dependent carboxylation of the biotinyl moiety on biotin carboxyl carrier protein (BCCP), and the carboxytransferase (CT) subunits catalyze the transfer of activated carboxyl groups from BCCP to acetyl CoA to form malonyl CoA.

The multifunctional (MF) ACCase, consisting of a single 220–240 kDa polypeptide with BCCP, BC, and CT domains, is nuclear encoded except the αCT subunit which is encoded by the plastidial genome [13]. In all plants, MF ACCase is involved in very long-chain fatty acid and flavonoid biosynthesis in the cytosol [13].

The sensitivity of plastidial ACCase to sethoxydim and the presence of a 220-kDa biotinylated polypeptide in soybean plastids provide a biochemical indication for the possible presence of two ACCase isoforms, one resistant (MS) and one sensitive (MF), in soybean leaf chloroplasts [14].

2.2 Fatty acid synthase (FAS)

The second enzyme complex involved in *de novo* synthesis is the fatty acid synthase (FAS). In nature, fatty acid synthases are subdivided into two groups. The fatty acid synthase type I which is characterized by a large, multifunctional proteins typical of yeast and mammals and the fatty acid synthase type II, found in prokaryotes which is composed of four dissociable proteins that catalyze individual reactions. Although plant cells are eukaryotic, the fatty acid synthase found in plastids is of type II [15]. Plant fatty acid synthase has inherited from photosynthetic prokaryotes; plastids being considered by the endosymbiotic theory as an old cyanobacteria. In plants, acyl carrier protein (ACP) is used as the acyl carrier for the various intermediate for fatty acid synthase unlike other eukaryotic cells where fatty acids are in acyl CoA form [15].

The initial substrates for fatty acid biosynthesis are acetyl CoA and malonyl-ACPs. The transfer of malonyl moiety from CoA to ACP is catalyzed by malonyl CoA:ACP transacylase (MAT). After the initial condensation of acetyl CoA and malonyl-ACP, all the intermediates for each step of the fatty acid biosynthetic pathway are acyl-ACPs.

AGS is composed of four enzymes: ketoacyl-ACP synthase (KAS, EC 2.3.1.41), β-ketoacyl-ACP reductase (EC 1.1.1.100), hydroxy acyl-ACP dehydrase (EC 4.2.1.59), and enoylacyl-ACP reductase (EC 1.3.1.9). All components of fatty acid synthase occur in plastids, although they are encoded in the nuclear genome and synthesized on cytosolic ribosomes. There are four sequential reactions involved in two-carbon addition (**Figure 2**).

The first condensation takes place between acetyl CoA and malonyl-ACP This reaction is catalyzed by 1,3-ketoacyl-ACP synthase III (KAS, EC 2.3.1.41), one of three ketoacyl synthases in plant systems [15]. KAS I is responsible for the condensations in each elongation cycle up through that producing palmitoyl-ACP (16:0-ACP). KAS II is dedicated to the final plastidial elongation, that of palmitoyl-ACP (16:0-ACP) to stearoyl-ACP (18:0-ACP).

The β-ketoacyl-ACP formed during the condensation reaction successively undergoes a reduction reaction by β-ketoacyl-ACP reductase (EC 1.1.1.100), dehydration by the β hydroxyacyl-ACP dehydratase (EC 4.2.1.59) and a further reduction by enoylacyl-ACP reductase (EC 1.3.1.9) to give butyryl-ACP. The coenzyme of the two oxidation-reduction reactions is NADPH (**Figure 3**). The butyryl-ACP formed

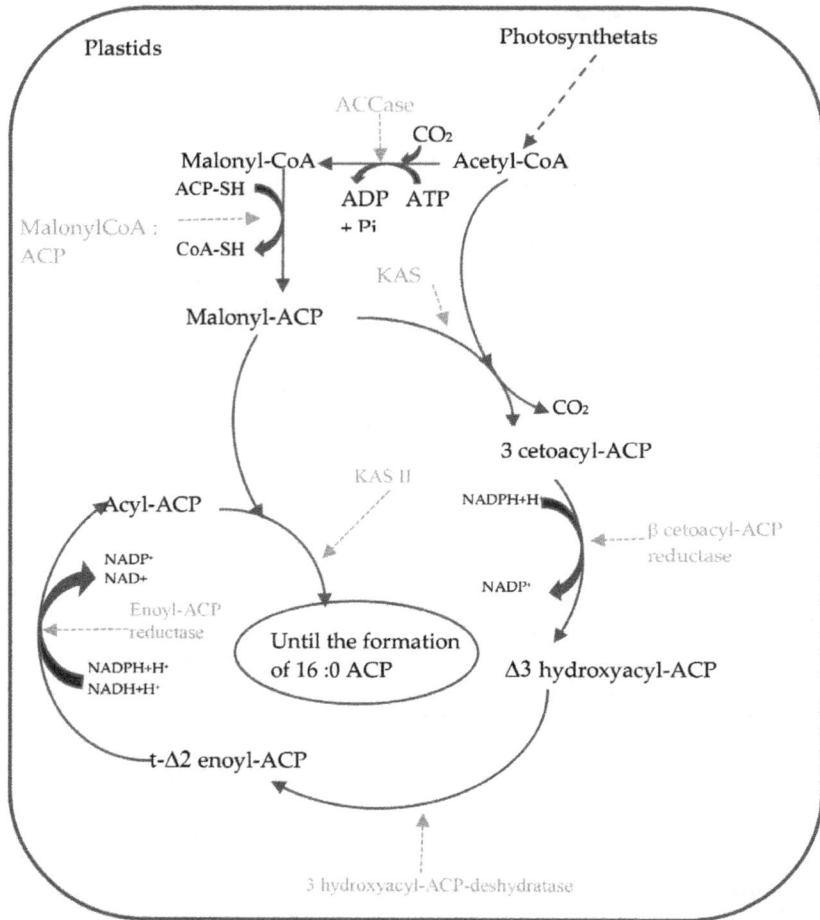

Figure 2.
Plant fatty acid biosynthesis. This chain requires a carboxylation reaction of acetyl CoA to malonyl CoA, an activation reaction of malonyl CoA to malonyl-ACP, a condensation reaction between acetyl CoA and malonyl-ACP to form β-ketoacyl-ACP, which undergoes in turn a reduction reaction, dehydration, and a second reduction extending the fatty acid of two carbon atoms.

will be extended by two further C2 units after further condensation with malonyl-ACP. The β-ketoacyl-ACP synthase I (KASI) catalyzes this reaction. After seven rounds of cycle, palmitoyl-ACP is formed.

Although the final product of fatty acid synthase is palmitic acid, two other common fatty acids are synthesized in the chloroplast stroma. These are stearic and oleic acids. The palmitoyl-ACP (C16:0-ACP) will be extended by two new units to form a stearoyl-ACP (C18:0-ACP) chain by a plastid soluble stearoyl-ACP synthase which is a multienzymatic complex composed of four enzymes (KASII, enoyl-ACP reductase, hydroxyacyl-ACP dehydrase, and enoylacyl-ACP reductase) [16].

The formed stearoyl-ACP is then desaturated with a plastidial soluble stearoyl-ACP desaturase (SAD, EC 1.14. 19.2) in oleoyl-ACP (C18:1Δ9-ACP) [17]. This enzyme is a nuclear-encoded, plastid-localized soluble desaturase that introduces the first Δ9 double bond into the saturated fatty acid resulting in the conversion of 18:0-ACP into 18:1Δ9-ACP [18].

The lack of structural similarity between plant and mammalian desaturase reflects the facts that the fatty acid substrates are on different carriers (ACP and

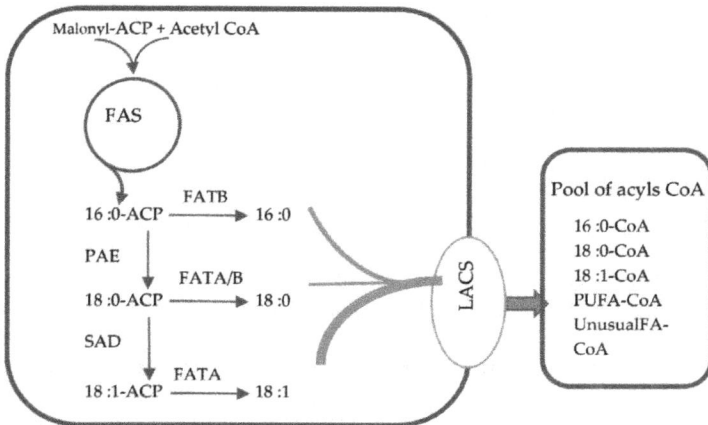

Figure 3.
Schematic representation of the export of fatty acids from the plastid to the cytosol. ACC, acetyl CoA carboxylase; ACP, acyl carrier protein; FA, fatty acid; CoA, coenzyme A; FAS, fatty acid synthase; FAT A/B, fatty acyl-ACP thioesterase A/B; LACS, long-chain acyl CoA synthase; PAE, palmitoyl-ACP elongase; SAD, stearoyl-ACP desaturase.

CoA, respectively),that the enzymes utilize different electron donors (ferredoxin vs. cytochrome b5), and that the plant enzyme is soluble, whereas the animal and fungal enzymes are integral membrane protein [19]. This enzyme plays a key role in determining the ratio of saturated to unsaturated fatty acids [17].

3. Glycerolipids as substrates for desaturation

In addition to the soluble acyl-ACP desaturases, the fatty acids synthesized in the chloroplast (palmitate, stearate, and oleate) are desaturated by membrane-bound desaturases that utilize complex lipid substrates such as phosphatidylcholine (PC) in the endoplasmic reticulum (ER) or monogalactosyl-diacylglycerol (MGDG) in the plastid [16]. These fatty acids were used for the synthesis of glycerolipids by two distinct metabolic pathways (prokaryotic and eukaryotic pathway) and in different cellular compartments (plastids and ER) [20].

The importance of both biosynthetic pathway depends on the plant species. In the "C18:3" plants photosynthetically active, only the extrachloroplast galactolipid pathway is functional; in the case of "C16:3" plants, the two pathways coexist and their importance differs according to the species and the conditions of the environment.

3.1 Importance of thioesterases

The flow of fatty acids (palmitoyl-ACP, oleoyl-ACP, and to a lesser extent stearoyl-ACP) through the two pathways would be subject to severe control. Two acyl-ACPs thioesterase enzymes and a chloroplast glycerol 3-phosphate acyl transferase play a very important role. Indeed, the acyl residues enter directly into the extrachloroplast pathway after having been hydrolyzed by fatty acyl-ACPs thioesterases FAT (A EC 3.1.2.14 and B EC 3.1.2.22) [22, 23] or in the chloroplast pathway after being acylated by acyl transferases [21].

The released palmitic, oleic, and stearic acids are then activated into coenzyme A ester by the action of long-chain acyl-CoA synthetase (LACS, EC 6.2.1.3) [24] and are exported to the cytosol (**Figure 3**).

Plants export sufficient fatty acid (16: 0-CoA, 18: 0-CoA, 18: 1-CoA) for lipid synthesis of extraplastid membranes and TAG of seed lipids of all plants.

3.2 Prokaryotic pathway

The prokaryotic pathway uses acyl-ACPs for PA and PG synthesis in all plants, and galactolipids (MGDG, DGDG, and SQDG) of so-called "C16:3" plants. This pathway is similar to the pathway demonstrated in photosynthetic prokaryotes [22].

The prokaryotic pathway is distinguished from the eukaryotic one by the presence of C16 fatty acids at the *sn*-2 position of the glycerol backbone. This pathway is characterized by the presence of molecular species 18: 3/16: 3 MGDG [23].

The major molecular species of MGDG synthesized by the prokaryotic pathway generally contain α-linolenic acid (C18:3), exclusively on the *sn*-1 position of glycerol backbone, while the *sn*-2 position is esterified by hexadecatrienoic acid (C16:3), resulting in desaturation of palmitic acid. The prokaryotic pathway, exclusively localized in plastids, therefore requires desaturation steps.

The DAG, precursor of prokaryotic MGDG, is an 18:1/16:0 DAG (**Figure 5**). The first molecular species synthesized by MGDG synthase is 18:1/16 0 MGDG. The palmitoyl residue is desaturated to a *cis*-hexadecenoyl residue by an ω9 desaturase, which is specific for both, the *sn*2 position of the fatty acid on glycerol and lipids (MGDG) [24, 25]. The ω9 desaturase is much more active on the palmitoyl residue in the *sn*2 position of glycerol of the MGDG than on the one located in the position *sn*2 of the DGDG.

The ω6 and ω3 desaturases, respectively, catalyze the desaturation of the monounsaturated acyls (hexadecenoyl and oleoyl) and diunsaturated (hexadecadienoyl and linoleoyl) residues. These desaturases have no specificity with respect to the length of the fatty acid chain or its position on glycerol. The ω6 desaturase acts equally well on the hexadecenoyl and oleoyl residues located at the *sn*2 and *sn*1 positions of the MGDGs and DGDGs. The desaturation of palmitic acid to hexadecenoyl acid is a prerequisite for other desaturations [26] (**Figure 4**).

This desaturation scheme is similar to that proposed for the desaturation of lipid acyls in the blue seaweed *Anabaena variabilis* [22].

Phosphatidyl glycerol synthesis occurs in both "C16:3" and "C18: 3" plants in the chloroplast. It involves in a first step the phosphatidic acid (PA) and CDP DAG which is of prokaryotic type. About 30–40% of the palmitoyl residue at position *sn*-2 of PG is desaturated at carbon 3 to form 3-*trans*-hexadecenoic acid [27]. The structure of this fatty acid is unusual; in plants, all the double bonds of the fatty acids of the membrane lipids are of cis type with the exception of this fatty acid.

3.3 Eukaryotic pathway

The second pathway, called the "eukaryotic" pathway, leads to the formation of two types of MGDG molecular species; one of them contains α-linoleate in both

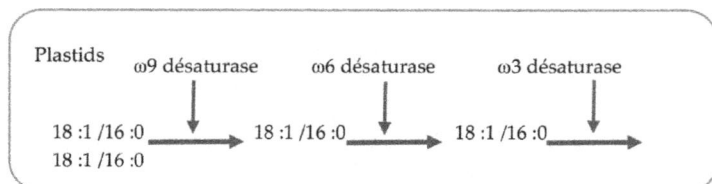

Figure 4.
Possible desaturation scheme of prokaryotic MGDG in plastids.

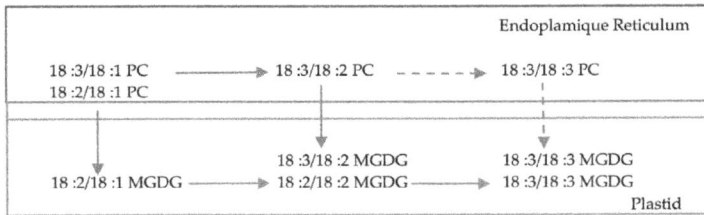

Figure 5.
Scheme of possible ways for the synthesis of eukaryotic-type MGDG in spinach, C16: 3 plants according to [25].

positions of glycerol and the other molecular species contains α-linoleate in position
sn-2 and palmitate in position *sn*-1. This pathway requires cooperation between
plastids and the endoplasmic reticulum for the formation of glyceroglycolipids in
chloroplasts [23, 28].

The oleate integrated into PC molecules at the position *sn*-2 of glycerol backbone
undergoes a succession of desaturations catalyzed by the (ω-6, Δ12) oleate desaturase, still identified by the fad2 mutation of *Arabidopsis* and allowing the synthesis
of linoleic acid and the (ω-3, Δ15) linoleate desaturase, identified by the fad3 mutation of *Arabidopsis*, which allows the synthesis of α-linoleic acid. Mutants deficient
for the lysophosphatidylcholine acyltransferase (LPCAT1 and LPCAT2 genes) have
reduced levels of polyunsaturated FA (PUFA) in TAGs [29].

After desaturation as acyl-PC, a part of them, probably in the form of DAG,
returns to the chloroplast and contributes to the formation of chloroplast galactolipids (**Figure 5**). These DAGs can be desaturated by chloroplast desaturases [30, 31].

It is therefore possible to judge the relative contributions of the prokaryotic and
eukaryotic pathways by comparing the proportions of eukaryotic 18/18 and 16/18
glycerolipids with prokaryotic 18/16 glycerolipids.

4. Synthesis of membrane glycerolipids

Membranes of eukaryotic cells have multiple functions in ensuring physical
compartmentalization at the cellular and subcellular levels, the regulation of
exchanges by the transport of metabolites and macromolecules, cellular communication (hormone receptors, surface antigens, signal transduction, etc.), and in
some specific metabolic reactions. In the same cell, it is therefore not surprising to
encounter different types of membranes with a specific lipid and protein composition that will determine their respective functions [32]. This synthesis is mainly carried out by two metabolic pathways described as prokaryotic and eukaryotic [33].

4.1 Synthesis of plastid lipids

Eukaryotic DAGs and prokaryotic DAG structures are the precursors of gly-
colipid synthesis (SQDG, MGDG, and DGDG). There are therefore two types of
glycolipids: prokaryotic glycolipids whose DAG backbone is of the C18/C16 type
and which are desaturated exclusively in the plastid and eukaryotic glycolipids
including DAGs of the type (C18:1/C18:1 and C16:0/C18:1) are derived from
phosphatidylcholine and are desaturated in RE and plastid [34]. The synthesis of
glycolipids, being localized in the membranes of the plastid envelope, thus requires
a mechanism for importing DAGs of eukaryotic structure. These differences in
DAG structure are due to different specificities of the chloroplast and ER acyl
transferases. The first step of the prokaryotic pathway is the transfer of the oleate to

a glycerol-3-phosphate at position *sn*-1 by an acyl ACP-glycerol 3 phosphate acyl-transferase (EC 2.3.1.1), soluble in the stroma of the plastid [35]. Lysophosphatidic acid (LPA) is thus formed. A second plastid-related plastid acyltransferase, the LPA-ACP acyltransferase, catalyzes the esterification of palmitoyl-ACP at the *sn*-2 position (LPAAT1; EC 2.3.1.51) [36]. This results in the synthesis of 18:1/16:0-PA.

Phosphatidic acid (PA) can either be converted to CDP-DAG by the action of a CTP-phosphatidate cytidylyltransferase (EC 2.7.7.41) which catalyzes the reaction between PA and CTP to form CDP-DAG and pyrophosphate or dephosphorylated to diacylglycerol (DAG) by phosphatidate phosphatase (PAP; EC 3.1.3.4). CDP-DAG will be used for the synthesis of phosphatidyl glycerol (PG) of the plastid [35] and DAG can be used for the synthesis of galactolipids (MGDG, DGDG) or a sulfolipid (sulfoquinovosyldiacylglycerol) (**Figure 6**).

From the phyllogenetic point of view, the difference between so-called "C16:3" and "C18:3" plants is related to the presence of plastid phosphatidate phosphatase in "C16: 3" plants, lost during evolution in "C18:3" plants. The chloroplast enzyme is clearly different from other phosphatidate phosphatases in the cell because it is membrane-bound, strongly associated with the inner membrane of the envelope, has an optimum alkaline pH, and is inhibited by cations such as Mg^{2+} [37]. The DAG thus produced (18/16 DAG) is at the origin of the glycolipids of prokaryotic struc-ture, SQDG, MGDG, and DGDG (**Figure 5**).

4.1.1 Synthesis of monogalactosyl diacylglycerol (MGDG)

MGDG is synthesized in a single step by a 1,2-DAG 3-β-galactosyltransferase (or MGDG synthase) that transfers galactose from UDP-Gal to DAG via a β1 → 3 glycosidic linkage [38]. MGDG synthase 1 catalyzes the synthesis of eukaryotic and prokaryotic MGDG molecules in vitro with no apparent specificity for either struc-ture [38] and is at the origin of the majority of the MGDG synthesized in standard

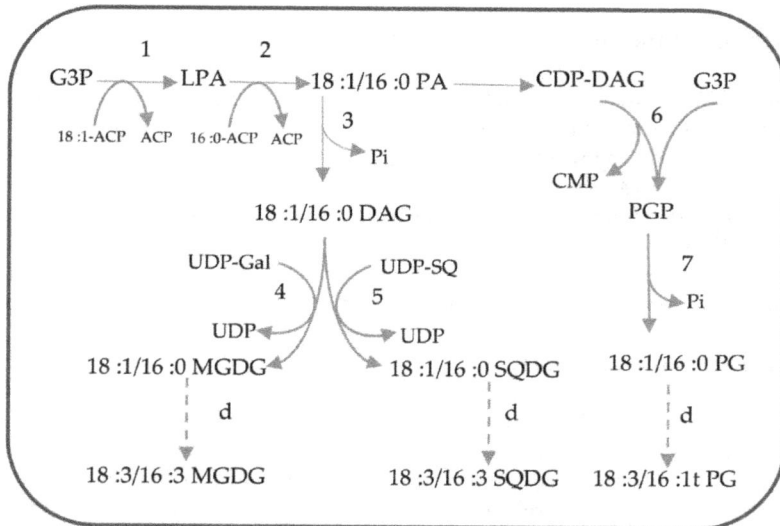

Figure 6.
Biosynthesis of glycerolipids according to the prokaryotic pathway (MGDG, DGDG, SQDG, and PG). The enzymes involved are: (1) glycerol-3-phosphate acyl transferase; (2) 1-acyl-glycerol-3-phosphoacyltransferase; (3) phosphatidate phosphatase; (4) MGDG synthase; (5) SQDG synthase; (6) phosphatidate cytidyl transferase; (7) CDP-DAG: glycerol-3-phospho-cytidylyltransferase; (8) phosphatidate glycerophosphatase; d: desaturase.

condition. In contrast, MGDG synthase 2 and 3 would be localized in the outer membrane [38]. These two enzymes have a better affinity for eukaryotic DAG (C18: 2/ C18: 2) [38] and would likely be in the supply of MGDG for synthesis of DGDG [39].

4.1.2 Synthesis of the DGDG

A small proportion of MGDGs are again glycosylated by DGDG synthase (EC 2.4.1.241) to form DGDG. Two enzymes catalyze DGDG synthesis by adding Gal from UDP-Gal to MGDG via α1 → 6 glycosidic linkage [40]. DGDG synthase1 acts preferentially on MGDG C18/C18, whereas DGDG synthase2 seems to have an affinity for MGDG with C16/18 [41]. These two enzymes would be localized in plastids, presumably in the outer membrane of the envelope [41].

4.1.3 Synthesis of sulfoquinovosyl-diacylglycerol SQDG

Similarly, a sulfolipid synthase (EC 3.13.1.1) catalyzes the attachment of UDP-sulfoquinovose (UDP-SQ) to the *sn*-3 position of DAG to form SQDG. The first step in the synthesis of SQDG or sulfolipid is the formation of UDP-SQ, a polar donor group [32]. The second reaction is catalyzed by a sulfolipid synthase (EC 3.13.1.1) that transfers SQ from UDP-SQ to a DAG molecule [42].

4.1.4 Synthesis of phosphatidylglycerol (PG)

Phosphatidic acid (PA) is also a substrate for CDP-DAG synthase (EC 2.7.7.41) to form CDP-DAG, the precursor of PG synthesis (**Figure 5**). In chloroplasts, PG is generated in the inner membrane of the envelope where phosphatidylglycerol-phosphate synthase and phosphatidylglycerol-phosphate phosphatase (EC 3.1.3.27) activities have been detected [43].

The fatty acids that make up the various glycerolipids formed in the plastid are characterized by a high degree of unsaturations introduced by the various fatty acid desaturases (FAD6, FAD7 and FAD8, EC 1.14.19) to generate polyunsaturated fatty acids (PUFA) necessary for the proper functioning of plastids [43].

4.2 Synthesis of glycerophospholipids in the endoplasmic reticulum

A major proportion of palmitic and oleic acids are transported as CoA esters outside the chloroplast to be incorporated at the endoplasmic reticulum (ER) into the phospholipids (PC, PE, PI, and PS) (**Figure 6**). ER is the main site for the synthesis of phospholipids and triacylglycerol, which derive from lysophosphatidic acid (LPA) as for the prokaryotic pathway (**Figure 7**).

In plant, the glycerophosphate acyltransferase (GPAT) family is involved in the first reaction leading to LPA synthesis of the eukaryotic pathway [35]. In the second reaction, cytosolic lysophosphatidic acid acyl transferase (LPAAT2, EC 2.3.1.23) specifically incorporates oleic acid at the *sn*-2 position of LPA, which is the specific signature glycerolipids from the eukaryotic pathway.

Most of the flow of chloroplast-exported fatty acids is incorporated in phosphatidylcholine (PC) by a mechanism called "acyl editing" [40]. This mechanism consists of a deacylation-reacylation cycle of the PC which makes it possible to exchange acyls present on the PC with activated FAs taken from a cytosolic pool of free acyl CoA. The oleate exported from plastids, in the form of oleoyl CoA, is used as a substrate for the synthesis of polyunsaturated fatty acids which are inserted either in membrane lipids (PC, PE, and PI) or in storage lipids (triacylglycerols TAG).

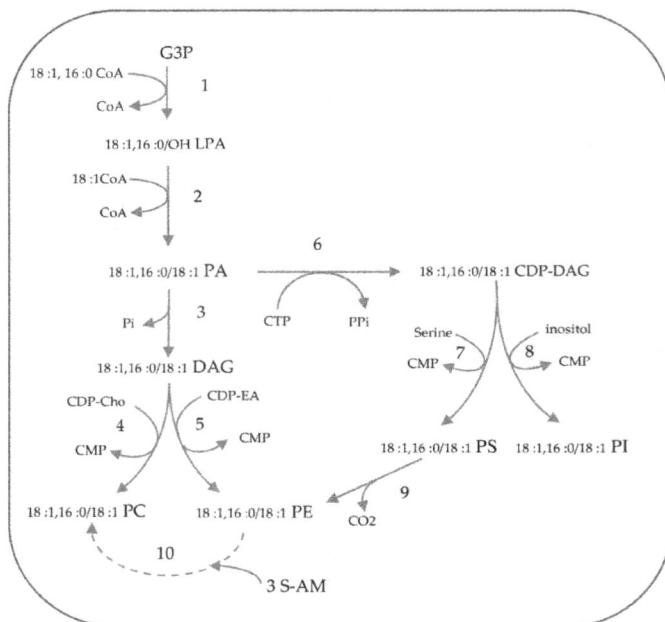

Figure 7.
Biosynthesis of glycerolipids according to the eukaryotic pathway (PC, PE, PI, and PS). The enzymes involved are: (1) glycerol-3-phosphate acyl transferase; (2) 1-acyl-glycerol-3-phosphoacyltransferase; (3) phosphatidate phosphatase; (4) CDP-choline: DAG choline phosphotransferase; (5) CDP-ethanolamine: DAG ethanolamine phosphotransferase or PE synthase; (6) phosphatidate cytidyl transferase; (7) PS synthase; (8) PI synthase; (9) PS decarboxylase; (10) N-methyltransferase.

In general, the synthesis of phospholipids is separated into three pathways: the phospholipids derived from cytidine diphosphate (CDP)-DAG (PI, PS), those derived from DAG (PC, PE) (**Figure 6**), and those from exchange of polar heads belonging to other phospholipids.

4.2.1 Phospholipids derived from CDP-DAG: PI and PS

PA can be converted to CDP-DAG by the action of a CTP phosphatidate cytidylyltransferase. This enzyme catalyzes the reaction between a eukaryotic PA molecule and a CTP molecule to form CDP-DAG and pyrophosphate.

Phosphoinositides are an important group of complex structure. PI represents 93% of phosphoinositides, while PIP (mainly PI-3P and PI-4P) and PIP2 (PI-(4,5) P2) represent less than 1%. These phosphoinositides play a major role in signaling processes. PI synthesis is catalyzed by PI synthase from free inositol and CDP-DAG. PI-3P and PI-4P are formed by phosphorylation of PI, respectively, by PI 3- and PI 4-kinases. Finally, PIP2 is formed from PI-4P by PI-4P 5-kinase activity (**Figure 7**).

PS synthase catalyzes the addition of serine to CDP-DAG [44].

4.2.2 Lipids derived from DAG: PE and PC

The plants synthesize ethanolamine by decarboxylation of serine [45], by serine decarboxylase which is a soluble, plant-specific enzyme. The synthesized free ethanolamine is then phosphorylated by an ethanolamine kinase, specific for ethanolamine different from choline kinase [46]. Phosphoethanolamine is then converted to CDP-ethanolamine by a CTP: phosphoethanolamine cytidyl transferase. The last step of PE synthesis is catalyzed by a CDP-ethanolamine: DAG ethanolamine phosphotransferase. This enzyme is an amino alcohol phosphotransferase that synthesizes both PE and PC [47] (**Figure 7**).

PC can also be synthesized by two different ways, either by methylation of PE with PE-N-methyltransferase, or by the addition of CDP-choline on DAG. The pathway using CDP-choline is preponderant [48].

The synthesis of all these lipids, phospholipids, and glycolipids is localized in specific membranes. However, a large part of the lipids thus generated is present in other membranes than those in which they are synthesized (vacuole, plasma membrane, and thylakoid). The cell therefore has specific lipid transport mechanisms.

5. TAG biosynthesis

TAGs are neutral lipids and are the major component of oilseed oil. These storage lipids represent the main source of carbon and energy mobilized during germination. Other tissues can also accumulate TAGs, such as senescence leaves or pollen grains [49, 50].

Their biosynthesis occurs at the ER membrane during the storage accumulation phase after embryogenesis. The TAGs result from the esterification at the *sn*-3 position of DAG of fatty acid from the pool of cytosolic acyl CoA by the action of diacylglycerol acyltransferase (DGAT, EC 2.3.1.20) or phospholipid: diacylglycerol acyltransferase (PDAT, EC 2.3.1.158) [17]. The acyl CoAs can be released from PC after desaturation with lyso-PC acyltransferase and reincorporated into the DAG-TAG chain. This allows for a renewal of the fatty acid composition of TAGs [51].

5.1 Seed triacylglycerols often contain unusual fatty acids

More than 300 different fatty acids are known to occur in seed TAG. Chain length may range from less than 8 to over 22 carbons. The position and number of double bonds may also be unusual, and hydroxy, epoxy, or other functional groups can modify the acyl chain.

The synthesis of these unusual fatty acids involves just one additional or alternative enzymatic step from primary lipid metabolism. All the enzymes identified to date that are involved in unusual fatty acid biosynthesis are structurally related to enzymes of primary lipid metabolism. Many of the unusual fatty acids are found in taxonomically dispersed families implying that the recruitment of enzymes for the synthesis of these unusual fatty acids might have occurred a number of independent times during angiosperm evolution.

Plants that synthesize **unusual monounsaturated fatty acids** have an additional desaturase, which is closely related to the Δ9-desaturase but introduces a double bond at a different location on the acyl-ACP. In coriander (*Coriandrum sativum*), petroselinic acid is synthesized by a desaturase that introduces a double bond between carbons 4 and 5 of a C16 acyl-ACP (Δ4-desaturase). This fatty acid is then extended by two carbons and cleaved from ACP to produce the free fatty acid. These last two steps are thought to require a specialized condensing enzyme and a specialized acyl-ACP thioesterase [52].

Plants that synthesize **medium-chain fatty** acids have several thioesterases. Indeed, plants that produce seeds with high concentrations of 8 to 14 carbon atoms, like *Cuphea lanceolata* rich in decanoic acid (C10: 0) *Umbellularia californica* rich in laurate (C12: 0) contain specific thioesterase for medium fatty acid chains. By removing acyl groups from ACP prematurely, the medium-chain thioesterases simultaneously prevent their further elongation and release them for triacylglycerol synthesis outside the plastids [53].

Seeds of *Ricinus communis* L. are the source of castor oil, used for the production of high-quality lubricants due to its high proportion of the unusual fatty acid ricinoleic acid. Castor bean seed oil contains 90% of the unusual hydroxy-fatty acid. Castor bean seeds contain an oleate hydroxylase which is structurally similar to extraplastidial membrane-bound $\Delta 12$-desaturases (FAD2), and only four amino acid substitutions are needed to convert an 18:1-desaturase into an 18:1-hydroxylase [54]. The synthesis of these fatty acids is thought to take place on the endoplasmic reticulum and use fatty acids esterified to the major membrane lipid phosphatidylcholine as a substrate.

Borage (*Borago officinalis* L.) seeds and evening primrose (*Oenothera biennis* L.) seeds are rich in γlinolenic acid ($\Delta 6, 9, 12$), respectively, from 22 to 25% and from 8 to 10%, an essential fatty acid. Its synthesis takes place in the RE during the formation of the seed. The precursor is a linoleoyl-PC and the desaturation is catalyzed by a D6 desaturase [55].

Very long-chain fatty acids (AGTLCs, containing more than 18 carbons) are used in the biosynthesis of many lipids involved in seed storage and waxes. Very long-chain fatty acids (VLCFAs) are synthesized in the following by-products of elongation of a C18 fatty acyl precursor by two carbons originating from malonyl CoA. Each elongation step requires four enzymatic reactions: condensation between an acyl precursor and malonyl-CoA, followed by a reduction, dehydration, and another reduction.

6. Conclusion

The reason for the great diversity in plant storage oils is unknown. The special physical or chemical properties of the "unusual" plant fatty acids have been exploited for centuries. Many of the unusual fatty acids have properties that are valuable as renewable feedstocks for the chemical industry. Medium fatty acids (lauric acid) are the ingredients of a soap or shampoo. VLCFAs like erucic acid (C22:1) can be used as a lubricant or participate in the formation of plastic film. Hydroxy fatty acids such as ricinoleic acid could be a source of biodiesel.

These unusual fatty acids synthesized by spontaneous plants are therefore obtained in small quantities. In order to obtain these fatty acids regularly and in large quantities for industrial use, it will either be necessary to domesticate the plant or introduce the specific gene of the nonconventional fatty acid into an oleaginous plant grown to obtain sufficient yields for industrial uses.

Author details

Fatiha AID
Laboratoire de Biologie et Physiologie des Organismes (LBPO), Université des Sciences et de la technologie Houari Boumediene (USTHB), Alger, Algerie

*Address all correspondence to: faid@usthb.dz

IntechOpen

References

[1] Gunstone FD, Harwood JL, Dijkstra AJ, editors. The Lipid Handbook. 3rd ed. Boca Raton, Florida: Taylor & Francis; 2007

[2] Podkowinski J, Jelenska J, Sirikhachornkit A, Zuther E, Haselkorn R, Gornicki P. Expression of cytosolic and plastid acetyl-coenzyme A carboxylase genes in young wheat plants. Plant Physiology. 2003;**131**:763-772

[3] Stumpf PK. The biosynthesis of saturated fatty acids. In: Stumpf PK, Cohn EE, editors. The Biochemistry of Plants. Vol. 9. New York: Acad. Press; 1987. pp. 121-136

[4] Harwood JL. Plant acyl lipids: Structure, distribution and analysis. In: Stumpf PK, Conn EE, editors. The Biochemistry of Plants. Vol. 4. New York: Academic Press; 1980. pp. 1-55

[5] Dubacq JP, Trémolières A. Occurrence and function of phosphatidylglycerol containing 3 *trans* hexadecenoic acid in photosynthetic lamellae. Physiologie végétale. 1983;**21**:293-312

[6] Dubacq JP, Drapier D, Trémolières A. Polyunsaturated fatty acid synthesis by a mixture of chloroplasts and microsomes from spinach leaves: Evidence of two distinct pathways of biosynthesis of trienoic acids. Plant and Cell Physiology. 1983;**24**:1-9

[7] Kinney AJ. Perspectives on the production of industrial oils in genetically engineered oilseeds. In: Kuo TM, Gardner HW, editors. Lipid Biotechnology. New York: Marcel Dekker; 2001. pp. 85-93

[8] Ohlrogge J, Browse J. Lipid biosynthesis. Plant Cell. 1995;**7**:957-970. DOI: 10.1105/tpc.7.7.957

[9] Lin M, Oliver DJ. The role of acetyl-coenzyme A synthetase in *Arabidopsis*. Plant Physiology. 2008;**147**:1822-1829

[10] Li-Beisson Y, Shorrosh B, Beisson F, Andersson MX, Arondel V, Bates PD, et al. Acyl-lipid metabolism. Arabidopsis eBook. 2010;**11**:01-65

[11] Roughan PG, Ohlrogge JB. Evidence that isolated chloroplasts contain an integrated lipid-synthesizing assembly that channels acetate into long-chain fatty acids. Plant Physiology. 1996;**110**:1239-1247

[12] Sasaki Y, Nagano Y. Plant acetylCoA carboxylase: Structure, biosynthesis, regulation and gene manipulation for plant breeding. Bioscience, Biotechnology, and Biochemistry. 2004;**68**(6):1175-1184

[13] Konishi T, Shinohara K, Yamada K, Sasaki Y. Acetyl-CoA carboxylase in higher plants: Most plants other than *Gramineae* have both the prokaryotic and the eukaryotic form of this enzyme. Plant and Cell Physiology. 1996;**37**:117-122

[14] Belkebir A, De Paepe R, Trémolières A, Aïd F, Benhassaine-Kesri G. Sethoxydim affects lipid synthesis and acetyl-CoA carboxylase activity in soybean. Journal of Experimental Botany. 2006;**57**:3553-3562

[15] Harwood JL. Fatty acid biosynthesis. In: Murphy DJ, editor. Plant Lipids: Biology, Utilisation and Manipulation. Oxford: Blackwell Publishing; 2005. pp. 27-66

[16] Harwood JL. Recent advances in the biosynthesis of plant fatty acids. Biochimica et Biophysica Acta. 1996;**1301**:7-56

[17] Zhang Y, Maximova SN, Guiltinan MJ. Characterization of a

stearoyl-acyl carrier protein desaturase gene family from chocolate tree, *Theobroma cacao* L. Frontiers in Plant Science. 2015;**6**:239. DOI: 10.3389/fpls.2015.00239

[18] Fox BG, Shanklin J, Somerville C, Munck E. Stearoylacyl carrier protein D9 desaturase from *Ricinus communis* is a di-iron-oxo protein. Proceedings of the National Academy of Sciences of the United States of America. 1993;**90**:2486-2490

[19] Shanklin J, Somerville C. Stearoyl-acyl-carrier-protein desaturase from higher plants is structurally unrelated to the animal and fungal homologs (fatty acid desaturation/cDNA clone/lipid unsaturation/fatty acid synthase). Proceedings of the National Academy of Sciences of the United States of America. 1991;**88**:2510-2514

[20] Roughan PG, Slack RC. Cellular organization of glycerolipid metabolism. Annual Review of Plant Physiology. 1982;**33**:97-132

[21] Heemskerk JWM, Wintermans JFGM. Role of the chloroplast in the leaf acid lipid synthesis. Physiologia Plantarum. 1987;**70**:558-568

[22] Löhden I, Frentzen M. Role of plastidial acyl-acyl carrier protein: Glycerol 3 phosphate acyltransferase and acyl-acyl carrier protein hydrolase in channeling the acyl flux through the prokaryotic and eukaryotic pathway. Planta. 1988;**176**:506-5012

[23] Sato N, Murata N. Lipid biosynthesis in blue green algae Anabaena variabilis. II Fatty acids and molecular species. Biochimica et Biophysica Acta. 1982;**710**:279-289

[24] Mazliak P, Justin AM, Demandre C, Chicha A. Specificity of some enzymes involved in glycerolipid biosynthesis. In: Cherif A et al., editors. Metabolism, Structure and Utilization of Plant Lipids. Tunis: Centre National Pedagogique; 1992. pp. 3-17

[25] Heemskerck JWM, Schmidt H, Hammer V, Wintermans JFGM. Biosynthesis and desaturation of prokaryotic galactolipids in and isolated chloroplasts from spinach leaves. Plant Physiology. 1991;**96**:144-152

[26] Ohlogge JB, Browse J, Somerville CR. The genetic of plant lipids. Biochimica et Biophysica Acta. 1991;**1082**:1-26

[27] Block MA, Douce R, Joyard J, Rolland N. Chloroplast envelope membranes: A dynamic interface between plastids and the cytosol. Photosynthesis Research. 2007;**92**:225-244

[28] Trémolières A, Dubacq JP, Drapier D, Muller M, Mazliak P. In vitro cooperation between plastids and microsomes in the leaf lipids. FEBS Letters. 1980;**114**:135-138

[29] Bates PD, Fatihi A, Snapp AR, Carlsson AS, Browse JA, Lu C. Acyl editing and headgroup exchange are the major mechanisms that direct polyunsaturated fatty acid flux into triacylglycerols. Plant Physiology. 2012;**160**:1530-1539

[30] Serghini-Caid H, Demandre C, Justin AM, Mazliak P. Oleoyl-PCmolecular species desaturated in pea leaf microsomes. Possible substrates of oleate desaturase in other green leaves. Plant Science. 1988;**54**:93-101

[31] Benning C. Mechanisms of lipid transport involved in organelle biogenesis in plant cells. Annual Review of Cell and Developmental Biology. 2009;**25**:71-91

[32] Wallis JG, Browse J. Mutants of *Arabidopsis* reveal many roles for membrane lipids. Progress in Lipid Research. 2002;**41**:254-278

[33] Shimojima M, Watanabe T, Madoka Y, Koizumi R, Yamamoto MP, Masuda K. Differential regulation of two types of monogalactosyldiacylglycerol synthase in membrane lipid remodeling under phosphate-limited conditions in sesame plants. Frontiers in Plant Science. 2013;**4**:469. DOI: 10.3389/fpls.2013.00469

[34] Xu C, Yu B, Cornish AJ, Froehlich JE, Benning C. Phosphatidylglycerol biosynthesis in chloroplasts of *Arabidopsis* mutants deficient in acyl-ACP glycerol-3phosphate acyltransferase. The Plant Journal. 2006;**47**:296-309

[35] Kim HU, Huang AHC. Plastid lysophosphatidyl acyltransferase is essential for embryo development in *Arabidopsis*. Plant Molecular Biology. 2004;**134**:1206-1216

[36] Block MA, Dorne AJ, Joyard J, Douce R. Preparation and characterization of membrane fractions enriched in outer and inner envelope membranes from spinach chloroplasts. II. Biochemical characterization. The Journal of Biological Chemistry. 1983;**258**:13281-13286

[37] Awai K, Maréchal E, Block MA, Brun D, Masuda T, Shimada H, et al. Two types of MGDG synthase genes, found widely in both 16:3 and 18:3 plants, differentially mediate galactolipid syntheses in photosynthetic and nonphotosynthetic tissues in *Arabidopsis thaliana*. Proceedings of the National Academy of Sciences of the United States of America. 2001;**98**:10960-10965

[38] Kobayashi K, Awai K, Takamiya K, Ohta H. *Arabidopsis* type B monogalactosyldiacylglycerol synthase genes are expressed during pollen tube growth and induced by phosphate starvation. Plant Physiology. 2004;**134**:640-648

[39] Kelly AA, Froehlich JE, Dörmann P. Disruption of the two digalactosyldiacylglycerol synthase genes DGD1 and DGD2 in *Arabidopsis* reveals the existence of an additional enzyme of galactolipid synthesis. Plant Cell. 2003;**15**:2694-2706

[40] Sanda S, Leustek T, Theisen MJ, Garavito RM, Benning C. Recombinant *Arabidopsis* SQD1 converts UDP-glucose and sulfite to the sulfolipid head group precursor UDP-sulfoquinovose in vitro. The Journal of Biological Chemistry. 2001;**276**:3941-3946

[41] Yu B, Xu C, Benning C. *Arabidopsis* disrupted in SQD2 encoding sulfolipid synthase is impaired in phosphate-limited growth. Proceedings of the National Academy of Sciences of the United States of America. 2002;**99**:5732-5737

[42] Gidda SK, Shockey JM, Rothstein SJ, Dyer JM, Mullen RT. *Arabidopsis thaliana* GPAT8 and GPAT9 are localized to the ER and possess distinct ER retrieval signals: Functional divergence of the dilysine ER retrieval motif in plant cells. Plant Physiology and Biochemistry. 2009;**47**:867-879

[43] Xu C, Härtel H, Wada H, Hagio M, Yu B, Eakin C, et al. The pgp1 mutant locus of *Arabidopsis* encodes a phosphatidylglycerolphosphate synthase with impaired activity. Plant Physiology. 2002;**129**:594-604

[44] Delhaize E, Hebb DM, Richards KD, Lin JM, Ryan PR, Gardner RC. Cloning and expression of a wheat (*Triticum aestivum* L.) phosphatidylserine synthase cDNA. Overexpression in plants alters the composition of phospholipids. The Journal of Biological Chemistry. 1999;**274**:7082-7088

[45] Rontein D, Rhodes D, Hanson AD. Evidence from engineering that decarboxylation of free serine is the major source of ethanolamine moieties

in plants. Plant and Cell Physiology. 2003;**44**:1185-1191

[46] Wharfe J, Harwood JL. Lipid metabolism in germinatingseeds. Purification of ethanolamine kinase from soya bean. Biochimica et Biophysica Acta. 1979;**575**:102-111

[47] Choi YH, Lee JK, Lee CH, Cho SH. cDNA cloning and expression of an aminoalcoholphosphotransferase isoform in Chinese cabbage. Plant and Cell Physiology. 2000;**41**:1080-1084

[48] Tasseva G, Richard L, Zachowski A. Regulation of phosphatidylcholine biosynthesis under saltstress involves choline kinase in *Arabidopsis thaliana*. FEBS Letters. 2004;**566**:115-120

[49] Kaup MT, Froese CD, Thompson JE. A role for diacylglycerol acyltransferase during leaf senescence. Plant Physiology. 2002;**129**:1616-1626

[50] Kim HU, Hsieh K, Ratnayake C, Huang AHC. A novel group of oleosins is present inside the pollen of *Arabidopsis*. The Journal of Biological Chemistry. 2002;**277**:22677-22684

[51] Battey JF, Schmid KM, Ohlrogge JB. Genetic engineering for plant oils: Potential and limitations. Trends in Biotechnology. 1989;**7**:122-125

[52] Grosbois M. Biosynthèse des acides gras au cours du développement du fruit et de la graine du lierre. Phytochemistry. 1971;**10**(6):1261-1273

[53] Pollard MA, Anderson L, Fan C, Hawkins DJ, Davies HM. A specific acyi-ACP thioesterase implicated in laurate production in immature cotyledons of *Umbellularia californica*. In: Quinn PJ, Harwood JL, editors. Plant Lipid Biochemistry, Structure and Utilization. London: Portland; 1990. pp. 163-165

[54] Van de Loo FJ, Turner PBS, Somerville C. An oleate 12 hydroxylase from *Ricinus communis* L. is a fatty acyl desaturase homolog (ricinoleic acid/castor/FAH12/transgenic plants). Proceedings of the National Academy of Sciences of the United States of America. 1995, 1995;**92**:6743-6747

[55] Galle AM, Demandre C, Guerche P, Joseph M, Dubacq JP, Mazliak P, et al. γlinolenic acid biosynthesis in microsomal membranes of developing borage officinal is seeds. In: Chérif A et al., editors. Metabolism, Structure and Utilization of Plant Lipids. Tunis: Centre National Pédagogique; 1992. pp. 185-188

Chapter 4

Effect of Nanoparticles on Lipid Peroxidation in Plants

Shahla Hashemi

Abstract

The size of the nanoparticles is between 1 and 100 nm. Nanoparticles are widely used in consumer and medical products, as well as in agricultural and industrial applications. The excessive use nanoparticles increases its release into the environment. Plants are an important part of the environment that is affected by nanoparticles. Studies have examined the effect of nanoparticles on plants. The results showed that high concentrations of nanoparticles showed a negative effect. Reactive oxygen species generation is a toxicological mechanism of nanoparticles in plants. When the production of radicals is greater than its removal, oxidative stress occurs. The key indicator of oxidative stress is lipid peroxidation. The unsaturated fatty acids in the cell membrane are a major target for radicals. Radical absorbs hydrogen from unsaturated fatty acids to form water. Therefore, the fatty acid has a non-coupled electron, which is then able to capture oxygen and form a peroxyl radical. Lipid peroxyl radical can lead to a chain of radical production. Enzymatic and nonenzymatic systems exist for the removal of radicals in plants. Enzymatic systems include catalase, guaiacol peroxidase, ascorbate peroxidase, superoxide dismutase, glutathione reductase, and dehydroascorbate reductase. Nonenzymatic systems include ascorbate and carotenoids, glutathione, tocopherol, and phenolic compounds.

Keywords: nanoparticles, reactive oxygen species, malondialdehyde, catalase, ascorbate, glutathione

1. Introduction

The nanoparticles have a size of less than 100 nm in at least one dimension. Due to the specific properties of nanoparticles, in particular the high-surface-to-volume ratio, they have been used for several applications. For example, nanoparticles are used in the fields of biosensors and electronics, cosmetic industries, wastewater treatment, biomedicines, cancer therapy, and targeted drug delivery [1, 2]. The excessive use of nanoparticles results in the release of these materials into the environment. The environment includes plants, the main producers of the food chain, which are affected by nanoparticles. Nanoparticles are absorbed by plants and transmitted to various parts of the plants and affect them. Several factors such as physicochemical properties of nanoparticles, plant species, and exposure conditions contribute to the absorption and transfer of nanoparticles. Size, magnetic properties, surface charge, composition, crystalline state, and surface functionalization are some of the physical properties of nanoparticles that are important in their absorption into the plant. Nanoparticles are introduced into the plant by various methods, for example, through penetration into the coating of seeds, during absorption of nutrient by the root, and entering the cuticle and stomata of the leaf. After absorbing nanoparticles, these materials can accumulate or move through the vascular system to the shoot. The first cell-level barrier to move nanoparticles is the cell wall. The size of the pores in the cell wall is 5–20 nm. Therefore, nanoparticles of less than 20 nm in size can easily pass through the pores. But nanoparticles with sizes larger than 20 nm through routes such as ion channels, endocytosis, and aquaporins and creation of new pores pass the cell wall of the barrier. The next barrier is the plasma membrane. The role of the plasma membrane is controlling the passage of materials in and out of the cell. Protein and lipids are two main parts of the plasma membrane structure. Plasma membrane lipids play an important role in determining cellular structures, regulating fluid membrane and signal transduction. Lipids are not only present in the plasma membrane but also in all parts of the plant. Plants have a diverse range of lipids including fatty acids, sterol lipids, glycolipids, sphingolipids, phospholipids, and waxes. In this chapter, we discussed the effects of nanoparticulate toxicity on the lipids of plants and plant defense mechanisms against this toxicity.

2. Interaction of nanoparticles with plants

There are reports that nanoparticles can lead to stress through release of reactive oxygen species (ROS) in plants. The lack of balance between the production and removal of ROS leads to the production of oxidative stresses with oxidative damage to DNA, proteins, and fats. There are two unpaired electrons in separate orbitals in the outer shell of oxygen. This oxygen structure makes it a candidate for the production of ROS. ROS are free radical species and non-free radical oxygen. Radicals can have neutral, negative, or positive charge. Free radical is an atom or group of atoms that have one or more unpaired electrons. Free radical oxygen species contains the hydroxyl radicals ($^{\bullet}OH$) and free radicals superoxide anion ($^{\bullet}O_2^-$). Non-free radical species containing hydrogen peroxide (H_2O_2) are various forms of activated oxygen resulted from oxidative biological reactions or exogenous factors (**Figure 1**). Radicals are naturally produced as intermediate biochemical reactions, but excess production of these radicals damages the plant and should be eliminated by the antioxidant system.

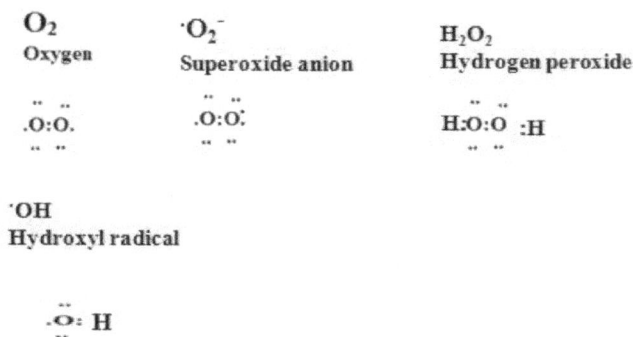

Figure 1.
Oxygen and some reactive oxygen species.

The antioxidant system in the plant contains an enzymatic and nonenzymatic system. The nonenzymatic antioxidant system contains alpha-tocopherol, flavonoids, ascorbate, glutathione and phenolic compounds, and carotenoids, while the enzymatic antioxidant system includes catalase (CAT), ascorbate peroxidase (APX), superoxide dismutase (SOD), peroxidase (POX), and glutathione reductase (GR).

3. Alpha-tocopherol

Alpha-tocopherol is a hydrophobic antioxidant that is produced by all plants (**Figure 2**). This compound is present primarily in the cell membrane and plays a key role in the collection of proxy lipid radicals from lipid peroxidation. One of the most prominent properties of tocopherol is their ability to turn off single oxygen, and it is estimated that a tocopherol molecule alone can neutralize about 120 molecules of single oxygen [3].

Alpha-tocopherol also acts as an end point for peroxidation reactions of unsaturated fats, which is converted to radical tocopheroxyl by reaction with lipid peroxyl radicals. Radical tocopheroxyl can be converted to tocopherol by reaction with ascorbic acid or other antioxidants [4] (**Figure 3**).

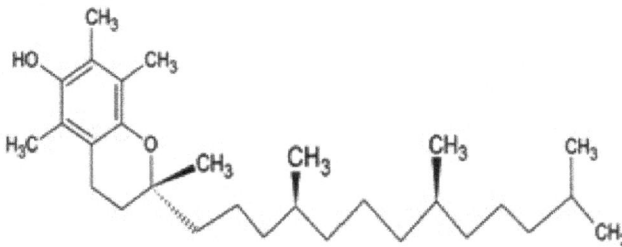

Figure 2.
Chemical structure of alpha-tocopherol [4].

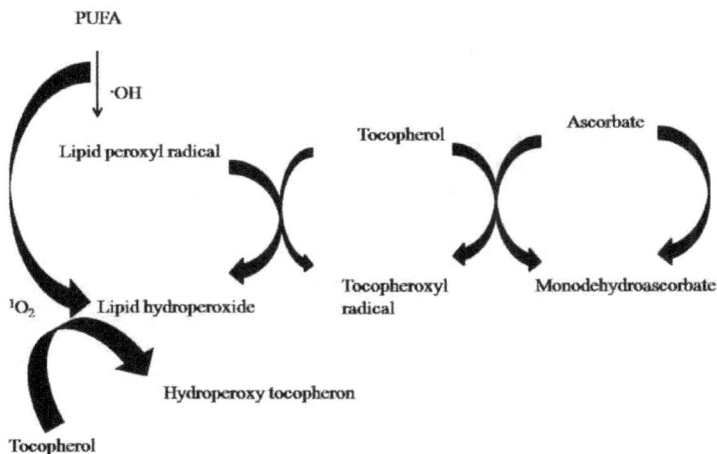

Figure 3.
The role of antioxidant tocopherol. Polyunsaturated fatty acids (PUFA).

4. Ascorbic acid

Ascorbic acid is one of the most powerful antioxidants that has been found in a variety of plant cells, organelles, and apoplastic space. In physiological conditions, ascorbic acid is often reduced. The ability of ascorbate to give the electron in a wide range of enzymatic and nonenzymatic reactions has transformed this substance into active oxygen species detoxification compound. Ascorbic acid plays a role in collecting superoxide, hydroxyl radicals, and singlet oxygen or converting hydrogen peroxide through the reaction of ascorbate peroxidase into water [5]. The conversion of hydrogen peroxide to water in ascorbate-glutathione cycle leads to the conversion of ascorbate to monodehydroascorbate. These compounds have a short life-span and convert into ascorbate by interfering with the enzymes of monodehydroascorbate reductase and NADPH (**Figure 4**).

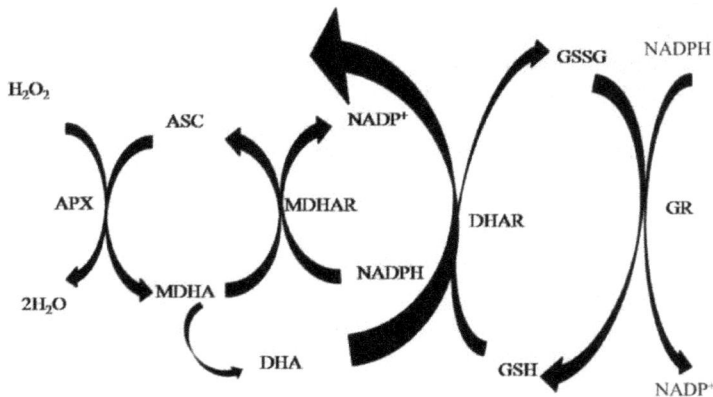

Figure 4.
Ascorbate-glutathione cycle. ASC, ascorbate; MDHA, monodehydroascorbate; DHA, dehydroascorbate; GSH, glutathione; GSSG, glutathione disulfide; APX, ascorbate peroxidase; MDHAR, monodehydroascorbate reductase; DHAR, dehydroascorbate reductase; GR, glutathione reductase [6].

5. Glutathione

Glutathione, γ-glutamyl-cysteinyl-glycine, is a tripeptide that has been found in all parts of the cell, such as cytosol, endoplasmic reticulum, chloroplast, vacuoles, and mitochondria [7]. Glutathione is one of the main sources of thiol in most plant cells. Due to the reactivity of the glutathione thiol group, this substance has been widely recognized for a wide range of biochemical reactions. The central cysteine in the glutathione molecule has created a high potential for reduction in this molecule. Reduced glutathione can remove the hydrogen peroxide [8]. The main role of glutathione in antioxidant defense is due to its ability to produce reduced ascorbic through the ascorbate-glutathione cycle. Some researchers have reported that glutathione protects the cell from oxidative stress by reacting thiol groups with singlet oxygen and radical hydroxyl [5, 9, 10].

6. Phenolic compounds

Phenolic compounds of a group of secondary metabolites include flavonoids and tannins, which are found in plant tissues abundantly. Most plants synthesize phenolic compounds in natural conditions, but their synthesis and accumulation are induced by stresses of nanoparticles [11]. Phenolic compounds play a role in H_2O_2-scavenging, quenching of singlet oxygen, and reducing or inhibiting lipid oxidation [14, 15]. There are two main mechanisms for the protective role of phenolic compounds (POH).

In the first mechanism, the hydrogen atoms of phenolic compounds are eliminated by free radical (R˙), and the phenolic compounds become radical:

$$R˙ + POH \rightarrow RH + PO˙$$

In evaluating the phenolic compounds' action in this mechanism, the bond dissociation energy of the O–H bonds is an important parameter, because the weakening of the OH bond increases the activity of phenolic compounds for radical deactivation [12].

In the second mechanism, free radicals can take electrons from phenolic compounds and convert them into radical cation [12]:

$$R˙ + POH \rightarrow R^- + POH˙^+$$

According to the second mechanism, the lower ionization potential of phenolic compounds releases electrons more easily.

Therefore, the activity of phenolic compounds is easily estimated by calculating ionization potential and the bond dissociation energy of the O–H bonds [12]. Bendary et al. suggested that the phenolic compounds perform scavenging of H_2O_2 from the first mechanism. The number of hydroxyl groups and the aromatic ring substitution pattern are all important associated factors. The ortho and para position substitution with another hydroxyl group is another important factor that plays in H_2O_2-scavenging [13].

7. Carotenoids

Carotenoids are tetraterpenes that exist in the photosynthetic and non-photosynthetic tissues of the plants and synthesize from isoprenoid biosynthesis pathway. Carotenoids act as auxiliary pigment in chloroplasts, but their main role is the role of antioxidant activity [14, 15].

There are two types of carotenoids in plant tissues:

1. Carotenoids that only contain hydrocarbons (carotene)

2. Carotenoids which in addition to the hydrocarbon chain have oxygen atoms (xanthophyll).

These compounds carry the antioxidant role through the following routes:

1. Eliminating singlet oxygen and wasting energy in the form of heat (**Figure 5**)

2. The reaction with excite chlorophyll and gaining energy to prevent the formation of singlet oxygen (**Figure 5**)

3. Waste high energy of exciting through the xanthophyll cycle [14, 15]

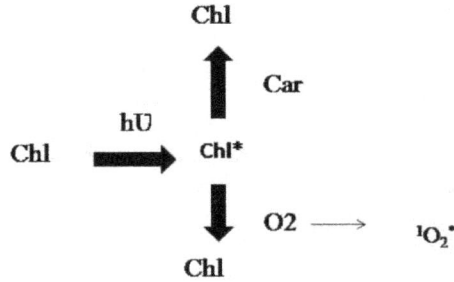

Figure 5.
*Schematic representation of singlet oxygen formation upon excitation of chlorophyll (Chl) and the role of carotenoids (car) in protection against photooxidative damage. The symbol * indicates excited states.*

8. The xanthophyll cycle

In green tissues, zeaxanthin epoxidase can be zeaxanthin converted to violaxanthin via the intermediate antheraxanthin. This is a reversible reaction; violaxanthin de-epoxidase converts violaxanthin to zeaxanthin by antheraxanthin. The relative concentration of zeaxanthin/violaxanthin is controlled by the xanthophyll cycle in plant photosynthetic tissues, which is a collection of light and dark control reactions.

Under high light conditions, violaxanthin de-epoxidase activated and converted violaxanthin to zeaxanthin. In dark conditions zeaxanthin is converted into violaxanthin [16] (**Figure 6**).

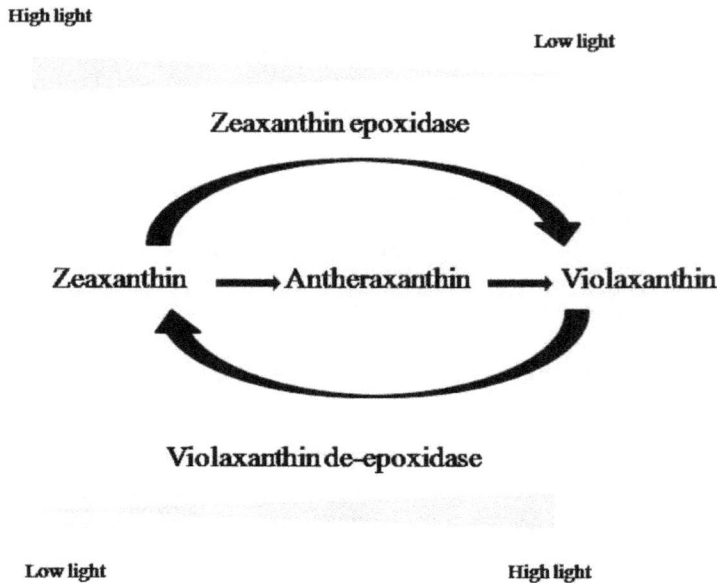

Figure 6.
The xanthophyll cycle.

9. Catalase

Catalase has a tetramer enzyme that breaks H_2O_2 into water and oxygen. Catalase has a lower affinity for H_2O_2, so it can remove H_2O_2 at high concentrations [17]. Hydrogen peroxide is very toxic to plant cells, especially in the chloroplast. Hydrogen peroxide at very low concentrations prevents the activity of the enzymes of the calvin cycle, especially the enzymes with *sulfhydryl* group such as glyceraldehyde 3-phosphate dehydrogenase and fructose 1,6 bisphosphatase [18, 19].

10. Peroxidases

Peroxides are a group of antioxidant enzymes that cause the decomposition of hydrogen peroxide with the oxidation of a substance. Peroxidases are located in the cytosol, vacuole, chloroplast, and extracellular space and are classified based on their combined composition.

11. Ascorbate peroxidase

Ascorbate peroxidase is an antioxidant enzyme that participates in the ascorbate-glutathione cycle, and its activity has been reported in the chloroplast, cytosol, peroxisome, and apoplast. This enzyme uses ascorbate as a reducing agent and decomposes hydrogen peroxide into water and oxygen [20]. The high concentration of ascorbate peroxidase to hydrogen peroxide shows that the ascorbate-glutathione cycle plays a vital role in controlling the level of radicals in cellular organs. In ascorbate-glutathione cycle with ascorbate peroxidase enzyme activity, ascorbate is oxidized to monodehydroascorbate, and ascorbate production is required to continue the cycle. In this cycle, the enzymes of monodehydroascorbate reductase (MADAR), dehydroascorbate reductase (DHAR), and glutathione reductase (GR) are active and reduce ascorbate using water and glutathione. Using NADPH, monodehydroascorbate reductase converts monodehydroascorbate to ascorbate. While dehydroascorbate reductase catalyzes dehydroascorbate to ascorbate using glutathione (GSH) oxidation.

12. Guaiacolperoxidase

Guaiacol peroxidase oxidize guaiacol. This enzyme is also present in the cytosol, vacuole, and cell wall [21].

13. Superoxide dismutase (SOD)

Superoxide dismutase is one of the enzymes that is located in all intracellular organs and apoplast and is very important in the defense against active oxygen species. This enzyme converts the radical superoxide to H_2O_2, which H_2O_2 should be detoxified during the next stages of antioxidant defense. In the presence of the superoxide dismutase enzyme, this reaction occurs 10,000 times faster [22].

$$O_2 \xrightarrow{e} \cdot O_2^- \xrightarrow{e} H_2O_2 \xrightarrow{e} \cdot OH \xrightarrow{e} H_2O$$

SOD

POX
CAT

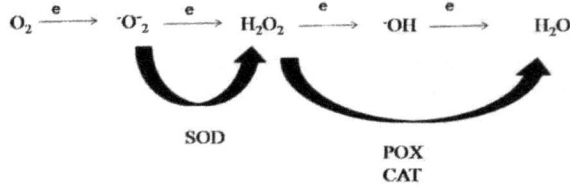

Figure 7.
Formation and elimination of reactive oxygen species.

This enzyme is divided into three groups based on its cofactor:

1. Cu/Zn SOD in the chloroplasts and cytosol

2. Fe/SOD in the chloroplasts of some plants

3. Mn/SOD in the mitochondrial matrix

There are genome types of SOD in the nucleus that are synthesized in the cytoplasm and then transmitted to different organs [23]. In the chloroplast, the superoxide dismutase enzyme is in two forms attached to the thylakoid membrane and is free in the stroma. The forms attached to the thylakoid membrane and the free enzyme, at the site of production and inward stroma, convert radical superoxide into H_2O_2 [24].

In summary, the activity of enzymes was shown in **Figure 7**.

14. Lipids and oxidation

Lipids are major constituents of prokaryotic and eukaryotic membranes. Besides serving as structural components of the plasma membrane and intracellular membranes, they provide diverse biological functions in energy and carbon storage, signal transduction, and stress responses. Plants contain a diverse set of lipids including fatty acids, phospholipids, glycolipids, sterol lipids, sphingolipids, and waxes. Polyunsaturated fatty acids (PUFAs) are lipid components commonly and easily oxidized by unbalanced ROS (mainly hydroxyl radical due to its indiscriminative reactive character).

15. Effect of reactive oxygen species on lipids

Cell membrane is one of the primary goals of many environmental stresses. Therefore, maintaining the integrity and stability of the membrane under stress is one of the signs of stress tolerance [25]. Polyunsaturated fatty acids are one of the most important membrane lipid compounds that are very sensitive to peroxidation. The main reason for the harmful effects of ROS is their ability to start the chain reaction of oxidation of unsaturated fatty acids, which leads to lipid peroxidation and membrane degradation.

The reaction is divided into three major steps: initiation, propagation, and termination.

Initiation

A radical fatty acid is produced at the initiation stage. Oxygen reactive species (ROS), such as OH, combines with hydrogen atom of unsaturated fatty acid to produce water and radical fatty acids (**Figure 8**).

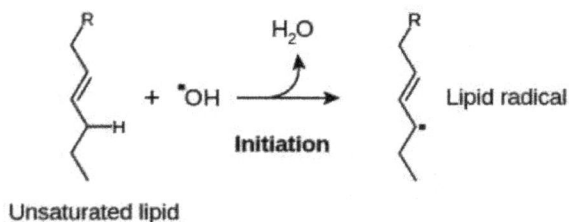

Figure 8.
The initiation phase of peroxidation of unsaturated fatty acids [26].

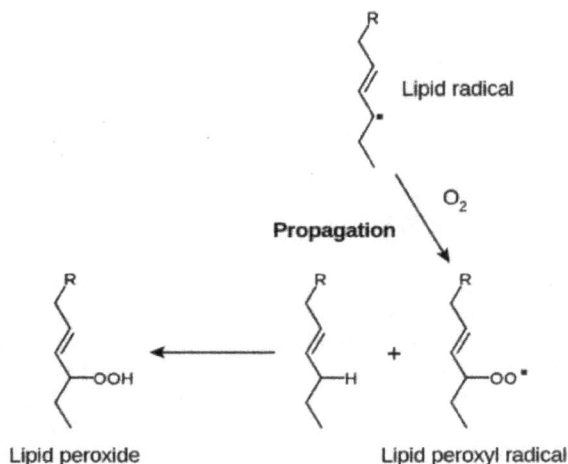

Figure 9.
The propagation phase of peroxidation of unsaturated fatty acids [26].

Propagation

The stability of the fatty acid radical is low, so that it reacts with molecular oxygen, resulting in the formation of peroxyl-fatty acid radical. This radical also has low stability, which reacts with another free fatty acid and produces a lipid peroxide and different radical fatty acid (**Figure 9**).

Termination

It always produces another radical, while a radical reacts with a non-radical. This process is called the "chain reaction mechanism." A non-radical species is produced when two radicals react together. So the radical reaction stops.

16. Formation malondialdehyde (MDA)

As previously mentioned, the polyunsaturated fatty acid (PUFA) acyl chain is attacked by free radicals and creates a radical fat that reacts easily with an oxygen molecule and forms a lipid peroxyl radical. The lipid peroxyl radical can attack neighboring PUFAs, propagating a chain reaction. A linolenic acid (18:3) peroxyl radical can also react internally, forming a cyclic peroxyl radical, which spontaneously reacts with a second oxygen molecule and is subsequently reduced to phytoprostane G1 (PPG1). Phytoprostane G1 (PPG1) either spontaneously decays, forming MDA and other alkanes and alkenes, or forms other phytoprostanes [27] (**Figure 10**).

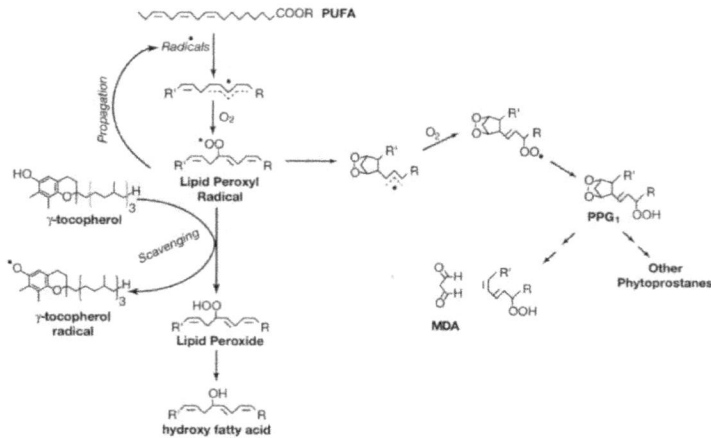

Figure 10.
Lipid peroxidation [27].

17. Malondialdehyde (MDA) assay

Lipid peroxidation was measured according to Heath and Packer (1968) by measuring the concentration of MDA. According to this method, 0.2 g of tissue was homogenized in 2 ml 0.1% (w:v) trichloroacetic acid (TCA) solution. The homogenate was centrifuged at 10,000 g (rcf) at 4°C for 10 min and 2 ml of supernatant transfer to new tube and then added 1 ml 20% TCA containing 0.5% (w:v) thiobarbituric acid (TBA). The reaction mixture was incubated in boiling water for 30 min at 95°C followed by placing the tubes on an ice bath to stop the reaction. The homogenate was centrifuged at 10,000 g for 15 min, and the absorbance was read at 532 nm [28]. The unspecific turbidity was corrected by A600 subtracting from A530. The amount of MDA–TBA complex (red pigment) was expressed as μmol/g FW and calculated by the extinction coefficient 155 mM − 1 cm − 1 using the formula (**Figure 11**).

$$MDA = \frac{(A532 - A600)}{155} \times \frac{Amount\ of\ reaction\ mixture}{tissue}$$

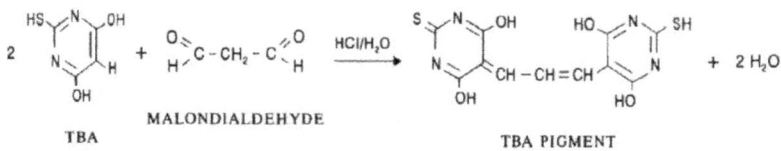

Figure 11.
The formation of MDA–TBA complex [29].

18. Radical scavenging assays

DPPH˙ is a stable free radical with purple color and a strong absorption band in the range of 515–520 nm. DPPH takes an electron or a hydrogen atom from an antioxidant accumulation molecule to become a stable DPPH molecule, in the presence of antioxidant compounds. The reduced form of DPPH is pale yellow. By studying spectrophotometric color changes, it may determine the antioxidant activity.

An antioxidant compound with a larger free radical scavenging capacity reduces DPPH further. Therefore, there is less purple color in the sample. The DPPH assay was performed according to the method developed by Blois (1958) slightly modified by Brand-Williams et al. [30, 31]. For 40 min, a solution of 1 mM DPPH$^•$ in 80% (v/v) methanol was stirred. Then a standard or sample (50 mL) was added with 2.95 mL of DPPH$^•$ solution and placed for 30 min in the dark.

Decrease of absorbance was read at 517 nm. DPPH scavenging effect is obtained from the following formula:

$$\text{DPPH scavenging effects}(\%) = \frac{A0 - A1}{A0} \times 100$$

A0: The absorbance is at 515 nm of the radical (DPPH) in the absence of antioxidant.

A1: The absorbance is at 515 nm of the radical (DPPH) in the presence of antioxidant.

19. ABTS$^+$ assay

Another important evaluation for antioxidant activity is the ABTS$^+$ test. In this assay, by peroxyl radicals or other oxidants, ABTS$^+$ is oxidized to its radical cation. ABTS$^{•+}$ is intensely colored (dark green). Reduction of color ABTS$^{•+}$ radical is used to measure the antioxidant capacity.

By using spectrophotometer, a decrease in absorbance by test compound and control is measured at 415 nm [32].

20. The role of enzymes in peroxidation of lipids

The lipoxygenase enzyme is one of the oxidative enzymes. This enzyme catalyzes the addition of molecular oxygen to unsaturated fatty acids, produces unsaturated fatty hydroperoxides, and accelerates lipid peroxidation. Free radicals produced by lipoxygenase cause irregularities in the selective membrane permeability. This irregularity leads to an increase in ion leakage, a decrease in the activity of ion pumps dependent on H$^+$ATPase, and changes in the cell membrane potential [33, 34].

21. Benefits of nanotechnology

Nanomaterials can offer many applications in mechanical industries especially in coating, lubricants, and adhesive applications. The magnetic nanoparticles such as Fe_3O_4 are employed in the biomedical and clinical fields. TiO_2 nanoparticles find an application in cosmetics, pigments, sunscreen products, solar cells, and photocatalysis [35]. However, human beings must take caution in using nanoparticles and nanotechnology.

Author details

Shahla Hashemi[1,2]

1 Biology Department, Faculty of Science, Shahid Bahonar University of Kerman, Iran

2 Young Researcher's Society, Shahid Bahonar University of Kerman, Kerman, Iran

*Address all correspondence to: shahlahashemi15@yahoo.com

IntechOpen

References

[1] Peralta-Videa JR, Zhao L, Lopez-Moreno ML, de la Rosa G, Hong J, Gardea-Torresdey JL. Nanomaterials and the environment: A review for the biennium 2008-2010. Journal of Hazardous Materials. 2011;**186**(1):1-15

[2] Nel A, Xia T, Mädler L, Li N. Toxic potential of materials at the nanolevel. Science. 2006;**311**(5761):622-627

[3] Yu S-W, Tang K-X. Map kinase cascades responding to environmental stress in plants. Acta Botanica Sinica. 2004;**46**(2):127-136

[4] Hashim S, Aisha AF, Majid AMSA, Ismail Z. A Validated Rp-Hplc Method for Quantification of Alpha-Tocopherol in *Elaeis guineensis* Leaf Extracts. 2013

[5] Sharma P, Dubey RS. Drought induces oxidative stress and enhances the activities of antioxidant enzymes in growing rice seedlings. Plant Growth Regulation. 2005;**46**(3):209-221

[6] Locato V, Cimini S, De Gara L. Strategies to increase vitamin C in plants: From plant defense perspective to food biofortification. Frontiers in Plant Science. 2013;**4**:152

[7] Mittler R. Oxidative stress, antioxidants and stress tolerance. Trends in Plant Science. 2002;7(9):405-410

[8] Larson RA. The antioxidants of higher plants. Phytochemistry. 1988;**27**(4):969-978

[9] Asada K. The water-water cycle in chloroplasts: Scavenging of active oxygens and dissipation of excess photons. Annual Review of Plant Biology. 1999;**50**(1):601-639

[10] Wang SY, Jiao H. Scavenging capacity of berry crops on superoxide radicals, hydrogen peroxide, hydroxyl radicals, and singlet oxygen. Journal of Agricultural and Food Chemistry. 2000;**48**(11):5677-5684

[11] Chung I-M, Rajakumar G, Thiruvengadam M. Effect of silver nanoparticles on phenolic compounds production and biological activities in hairy root cultures of *Cucumis anguria*. Acta Biologica Hungarica. 2018;**69**(1):97-109

[12] Bendary E, Francis R, Ali H, Sarwat M, El Hady S. Antioxidant and structure–activity relationships (Sars) of some phenolic and anilines compounds. Annals of Agricultural Science. 2013;**58**(2):173-181

[13] Ma X, Li H, Dong J, Qian W. Determination of hydrogen peroxide scavenging activity of phenolic acids by employing gold nanoshells precursor composites as nanoprobes. Food Chemistry. 2011;**126**(2):698-704

[14] Sun T, Yuan H, Cao H, Yazdani M, Tadmor Y, Li L. Carotenoid metabolism in plants: The role of plastids. Molecular Plant. 2018;**11**(1):58-74

[15] Nisar N, Li L, Lu S, Khin NC, Pogson BJ. Carotenoid metabolism in plants. Molecular Plant. 2015;**8**(1):68-82

[16] Demmig-Adams B, Adams WW. Antioxidants in photosynthesis and human nutrition. Science. 2002;**298**(5601):2149-2153

[17] Wang W-B, Kim Y-H, Lee H-S, Kim K-Y, Deng X-P, Kwak S-S. Analysis of antioxidant enzyme activity during germination of alfalfa under salt and drought stresses. Plant Physiology and Biochemistry. 2009;**47**(7):570-577

[18] Pan Y, Wu LJ, Yu ZL. Effect of salt and drought stress on antioxidant enzymes activities and SOD isoenzymes of liquorice (Glycyrrhiza uralensis

fisch). Plant Growth Regulation. 2006;**49**(2-3):157-165

[19] Takeda T, Yokota A, Shigeoka S. Resistance of photosynthesis to hydrogen peroxide in algae. Plant and Cell Physiology. 1995;**36**(6):1089-1095

[20] Sgherri CLM, Maffei M, Navari-Izzo F. Antioxidative enzymes in wheat subjected to increasing water deficit and rewatering. Journal of Plant Physiology. 2000;**157**(3):273-279

[21] Jaleel CA, Riadh K, Gopi R, Manivannan P, Ines J, Al-Juburi HJ, et al. Antioxidant defense responses: Physiological plasticity in higher plants under abiotic constraints. Acta Physiologiae Plantarum. 2009;**31**(3):427-436

[22] Blokhina O, Virolainen E, Fagerstedt KV. Antioxidants, oxidative damage and oxygen deprivation stress: A review. Annals of Botany. 2003;**91**(2):179-194

[23] Weisiger RA, Fridovich I. Mitochondrial superoxide dismutase site of synthesis and intramitochondrial localization. Journal of Biological Chemistry. 1973;**248**(13):4793-4796

[24] Jithesh M, Prashanth S, Sivaprakash K, Parida AK. Antioxidative response mechanisms in halophytes: Their role in stress defence. Journal of Genetics. 2006;**85**(3):237

[25] Sudhakar C, Lakshmi A, Giridarakumar S. Changes in the antioxidant enzyme efficacy in two high yielding genotypes of mulberry (*Morus alba* L.) under Nacl salinity. Plant Science. 2001;**161**(3):613-619

[26] Sachdeva M, Karan M, Singh T, Dhingra S. Oxidants and antioxidants in complementary and alternative medicine: A review. Spatula DD. 2014;**4**(1):1-16

[27] Sattler SE, Mène-Saffrané L, Farmer EE, Krischke M, Mueller MJ, DellaPenna D. Nonenzymatic lipid peroxidation reprograms gene expression and activates defense markers in arabidopsis tocopherol-deficient mutants. The Plant Cell. 2006;**18**(12):3706-3720

[28] Heath RL, Packer L. Photoperoxidation in isolated chloroplasts: I. kinetics and stoichiometry of fatty acid peroxidation. Archives of Biochemistry and Biophysics. 1968;**125**(1):189-198

[29] Shafel T, Lee SH, Jun S. Food preservation technology at subzero temperatures: A review. Journal of Biosystems Engineering. 2015;**40**(3):261-270

[30] Blois MS. Antioxidant determinations by the use of a stable free radical. Nature. 1958;**181**(4617):1199

[31] Brand-Williams W, Cuvelier M-E, Berset C. Use of a free radical method to evaluate antioxidant activity. LWT- Food Science and Technology. 1995;**28**(1):25-30

[32] Re R, Pellegrini N, Proteggente A, Pannala A, Yang M, Rice-Evans C. Antioxidant activity applying an improved Abts radical cation decolorization assay. Free Radical Biology and Medicine. 1999;**26**(9-10):1231-1237

[33] Kubi J. Exogenous spermidine alters in different way membrane permeability and lipid peroxidation in water stressed barley leaves. Acta Physiologiae Plantarum. 2006;**28**(1):27-33

[34] Juan M, Rivero RM, Romero L, Ruiz JM. Evaluation of some nutritional and biochemical indicators in selecting salt-resistant tomato cultivars. Environmental and Experimental Botany. 2005;**54**(3):193-201

[35] Hashemi S, Asrar Z, Pourseyedi S, Nadernejad N. Investigation of Zno nanoparticles on Proline, anthocyanin contents and photosynthetic pigments and lipid peroxidation in the soybean. IET Nanobiotechnology. 2018;**13**(1):66-70

Chapter 5

Fatty Acid Compositions in Fermented Fish Products

Afnan Freije and Aysha Mohamed Alkaabi

Abstract

Fish fermentation differs from one region to another in the world. Different types of fish, different fermentation conditions, and different fermentation processes are used, thus resulting in different fermented fish products. The most investigated fermented fish products in regard to the fatty acid contents are Kejeik from Sudan, Fseekh from Egypt, Hatahata-zushi from Japan, and Tareeh and Mehiawah from the Middle East. The results of those studies were not consistent regarding the effect of the fermentation process on the contents of saturated fatty acids (SFAs), monounsaturated fatty acids (MUFAs), and polyunsaturated fatty acids (PUFAs). Some of those studies reported an increase in the level of SFAs and a decrease in the PUFAs contents, while other studies reported the opposite. The fermentation process itself was attributed to different microorganisms such as lactic acid bacteria (LAB), halophilic bacteria, the bacterial flora of *Micrococcus* and *Bacillus* species, and a new bacillus strain named *Bacillus mojavensis-ASK*. Autolytic enzymes from the fish were also reported to be responsible for the fermentation process.

Keywords: fatty acids, fish, fermentation, PUFAs

1. Introduction

Lipids are considered as one major group of the naturally occurring organic molecules, along with carbohydrates, proteins, and nucleic acids [1]. Lipids are characterized according to their solubility (physical property) rather than their structure, in which they are insoluble in water, but soluble in nonpolar organic solvents, such as chloroform and benzene [1, 2]. All lipids are composed of carbon, hydrogen, and oxygen atoms; however, some lipids contain phosphorus, sulfur, nitrogen, or other elements [3]. Lipids are divided into three classes, (a) triacylglycerol's (TGs), which are used as long-term energy stores such as fats and oils; (b) phospholipids (PLs), which function primarily in cell membranes; and (c) steroids, like cholesterol which is a component of animal cell membranes and a precursor in the synthesis of various steroid hormones [1, 3]. Lipids play a structural role in the cell membranes in combination with proteins to give the membranes their semipermeable property [1]. In addition, lipids give the membranes their shape and protect them from the external environment [3].

1.1 Fatty acids (FAs)

Fatty acids are the building blocks for the majority of lipids especially TGs and PLs [3, 4]. FAs are composed of a long hydrocarbon chain (nonpolar) that is conjugated to a carboxyl group (polar) which is an acidic functional group [5]. FAs hydrocarbon chains range in length from 2 to 80 but commonly from 12 up to 24. Chain length from 2 to 6 are called short-chain, from 6 to 10 are called medium-chain, and 12 up to 24 are called long-chain [6]. FAs containing 16 and 18 carbons are the most prevalent. The majority of FAs have an even number of carbon atoms because they are synthesized by combining the C_2 units of acetyl CoA [7, 8]. FAs are usually synthesized by the enzyme fatty acid synthase that is responsible to convert acetyl CoA into fatty acid [5]. The hydrocarbon chains of FAs are usually unbranched and can be divided into saturated and unsaturated [9]. Saturated fatty acids (SFAs) have no double bond in their hydrocarbon chain, while unsaturated fatty acids (UFAs) have one or more double bonds in their chain. These double bonds cause the formation of bents or "kinks" in the fatty acid chains making them liquid at room temperature [3].

UFAs are divided into monounsaturated fatty acids (MUFAs) which have one double bond in their chain and polyunsaturated fatty acids (PUFAs) which have two or more double bonds [10]. The double bonds locations follow a unique pattern; the MUFAs usually have the double bond between carbons 9 and 10 (Δ^9) where C1 is the carboxyl carbon. The second double bond in PUFAs is mostly between carbons 12 and 13 (Δ^{12}). PUFAs do not normally have conjugated double bonds (—CH=CH—CH=CH—), instead their double bonds are usually separated by at least one methylene group (—CH=CH—CH2—CH=CH—) [7–10]. The stereochemistry of the double bond in the naturally occurring UFAs is usually *cis*, i.e., the two hydrogens on the carbon atoms of the double bond are on the same side of the molecule [9, 10]. *Trans* FAs are formed during the hydrogenation process of UFAs to produce SFAs [2]. In the *trans* isomers, the two hydrogens on the carbon atoms of the double bond are on the opposite sides of the molecule [10].

1.2 Nomenclature of fatty acids

Fatty acids can be named using (a) systematic naming addressed in detail by the International Union of Pure and Applied Chemists and the International Union of Biochemistry and Molecular Biology (IUPAC-IUBMB) and (b) common naming [4]. The systematic naming of FAs ends with the suffix "oic acid" on to the name of the parent hydrocarbon. However, the ionized form of the fatty acid at physiological pH ends with the suffix "ate" rather than "oic acid." FAs are named according to the total number of carbon atoms, in addition to the number and position of double bonds, if present. The carbon atoms in FAs are numbered from the carboxylic acid residue (C1), and the position of the double bond is described by the symbol delta (Δ) followed by the number of the first carbon involved in the bond. For example, the full systematic name for palmitoleic acid (common name) is *cis*-Δ^9hexadecenoic acid, and it is written as 16:1(Δ^9) in which it consists of 16 carbons in its hydrocarbon chain with a double bond positioned between carbons 9 and 10 [10].

The common names for FAs reflect their prominent food sources such as palmitic acid (C16:0) found in palm oil and *arachidonic* acid (*cis*5, *cis*8, *cis*11, *cis*14–20:4; from *arachis*, meaning legume or peanut) found in peanut butter [11]. The common names are much simpler than the systematic names; however, they do not give any information about the structure of the fatty acids.

Omega (ω, n) system is an alternative system of naming fatty acids. In this system, carbon atoms are numbered relative to the methyl end of the molecule. Omega system can be distinguished from the IUPAC naming system in the bases of the following: (a) omega nomenclature is only applied to unsaturated fatty acids, (b) omega system does not identify whether the double bond have *cis* or *trans* configuration, and (c) omega system does not identify the location of other double bonds in the molecule. For example, linoleic acid C18:2n-6($\Delta^{9,12}$) can be written as ω-6 fatty acid in omega system [11].

1.3 Essential fatty acids (EFAs)

Fatty acids that the body needs but cannot synthesize in sufficient amounts to meet physiological needs due to the absence of the required enzymes are called essential fatty acids. The body cannot synthesize two polyunsaturated fatty acids: linoleic acid (C18:2n6, LA) and alpha-linolenic acid (C18:3n-3, ALA); therefore, they must be supplied by the diet [12]. Animal cells are unable to introduce double bonds in the n-3 and n-6 positions because they are deficient in certain required desaturase enzymes. However, these cells have the ability to introduce double bonds into all other positions in fatty acid hydrocarbon chain [13]. Dietary EFAs are used to produce long-chain polyunsaturated fatty acids [14].

1.3.1 Omega-3 and omega-6 fatty acids

Although omega-6 FAs were the first to be described as an essential fatty acids in the 1920s, omega-3 FAs take more attention than omega-6 [13]. Linolenic acid (ω-3) is the parent of the omega-3 FA family in which it can be used to produce other members of the family (**Figure 1**) [12, 15].

Alpha-linolenic acid is the precursor for linolenic acid that is used to synthesize two active biological components: eicosapentaenoic acid (C20:5n-3, EPA) and docosahexaenoic acid (C22:6n-3, DHA) [12, 14, 16]. ALA is found in plant oils, nuts, and seeds such as flaxseeds, walnuts, soybeans [12].

Aquatic ecosystems are the principal sources of DHA and EPA in the biosphere provided by the fish and fish oils in human diet [12, 16]. On the other hand, linoleic acid can be used to produce other members of the omega-6 family like arachidonic acid (C20:4n-6, AA) that acts as a starting material for a number of eicosanoids, i.e., biologically active molecules that regulate body functions [12].

Omega-6 FAs are found in seeds, nuts, and vegetable oils of corn, sesame, and sunflower [12]. Before the industrialization, the ratio of omega-6 to omega-3 (n-6/n-3) was around 1:1 to 2:1 due the abundant consumption of vegetables and seafood high in omega-3 fatty acids. However, there is a gradual change in this ratio with the industrialization, mainly due to the consumption of refined oils and seeds with a high content of omega-6 fatty acids in which the (n-6/n-3) average ratio has been around 10:1 to 20:1 [17].

This imbalanced n-6/n-3 ratio can even be worsened by the overnutrition habit associated with the Western diet around the world. Overnutrition, associated with an increased amount of FAs made available for oxidation in the liver, favors the pro-inflammatory state due to the depletion of n-3 long-chain PUFA (n-3 LCPUFA) such as EPA and DHA, elevated n-6/n-3 ratio, hyperinsulinemia, and insulin resistance (IR). Such changes may result in the development of nonalcoholic fatty liver disease (NAFLD), steatosis [hepatocyte triacylglycerol accumulation, and cirrhosis (steatohepatitis)] [18].

1.3.2 The role of omega-3 and omega-6 fatty acids

The consumption of omega-3, omega-6 PUFAs, and their derivatives has various beneficial effects, ranging from fetal development to cancer prevention [19]. PUFAs have a preventive effect against arterial hypertension, asthma, inflammatory diseases, human breast cancer, and disorders of the immune system [20].

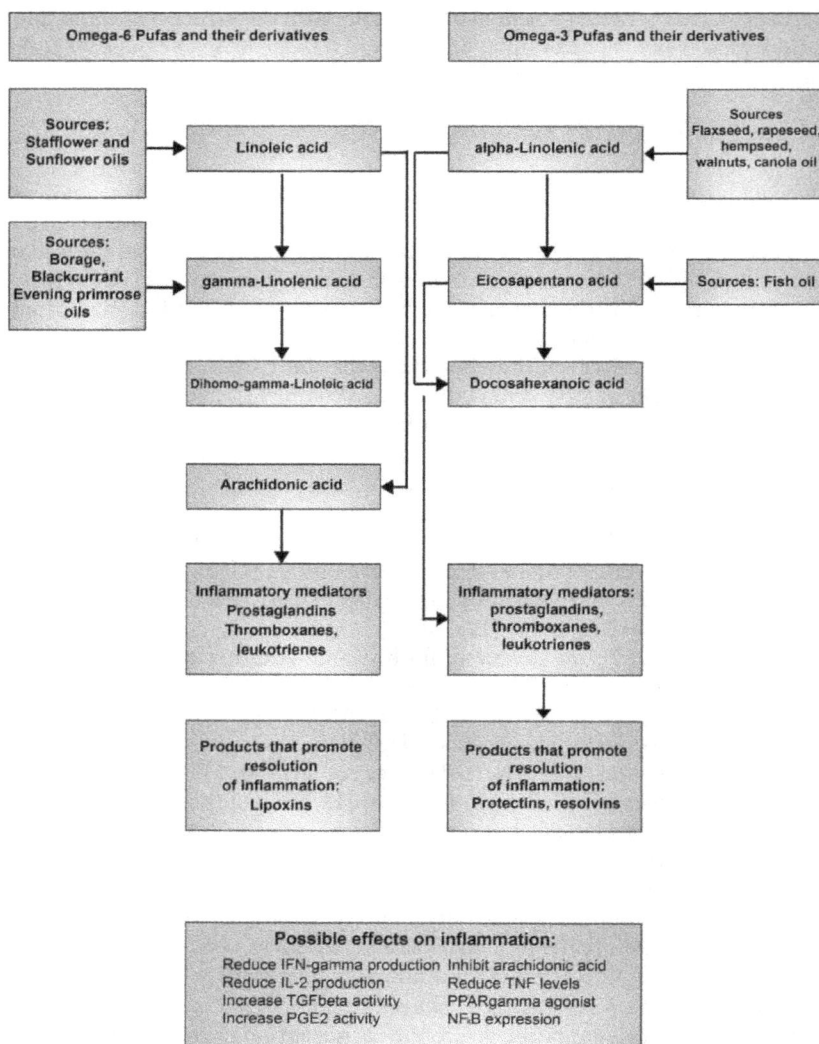

Figure 1.
Omega-6 and omega-3 PUFAs and their respective sources and metabolic derivatives [15].

n-3 FAs protect against several types of cardiovascular diseases such as myocardial infarction, atherosclerosis, arrhythmia, hypertension, and human coronary artery disease [19, 21]. Moreover, consumption of n-3 FAs interferes with prostaglandin metabolism, which can reduce the platelet aggregation and adhesion to blood vessels, which reduce the blood pressure [22]. Omega-3 PUFAs can reduce the incidence of high cholesterol level in the blood, psoriasis, cancer, and arthritis [23].

The n-3 LCPUFA DHA plays a major role in the development of the nervous system during the life of fetus and neonates [24]. The study of Barrera et al. (2018) has shown that a low dietary intake of n-3 LCPUFA in pregnant Chilean women has resulted in a significant decline in their erythrocytes and breast milk DHA levels. Therefore, the improvement of the quality of FAs intake specifically DHA was recommended during pregnancy and lactation periods in order to supply adequate amount of DHA to embryos and neonates [25]. In addition, n-3 supplements intake was suggested to improve the low level of n-3 LCPUFA associated with NAFLD patients in order to lower their n-6/n-3 ratio and thus reduce the inflammatory status as a treatment for the nutritional hepatic steatosis in adults [24].

Omega-6 PUFAs are converted to other important compounds through various enzymatic reactions to the key intermediate arachidonic acid, which is subsequently converted to eicosanoids that act somewhat like hormones such as prostaglandins, thromboxanes, and leukotrienes [15]. Eicosanoids play an important role in muscle relaxation and contraction, blood clot formation, blood vessel contraction and dilution, blood lipid regulation, and immune response to injury and infections [12].

Omega-3 PUFAs are considered as more potent anti-inflammatory agents than n-6 PUFAs, but the effects of n-6 PUFAs are more dominant due to the abundance of these compounds in the diet [15].

1.3.3 EPA and DHA

Alpha-linolenic acid can be used in the body as a precursor to produce EPA; however, EPA can be directly obtained through the consumption of fish oils. Moreover, EPA can be converted to DHA and also to eicosanoids that are essential for inflammatory signaling, such as prostaglandins, thromboxanes, and leukotrienes. DHA can also be converted to eicosanoids to produce anti-inflammatory mediators such as protectins and resolvins [15]. In 2004, The International Society for the Study of Fatty Acids and Lipids (ISSFAL) has recommended the minimum intake of 0.5 g/day of EPA and DHA for the prevention of cardiovascular disease [26].

1.4 Metabolism of long-chain PUFAs

ALA and LA obtained from diet can be converted into longer chains of PUFAs by the help of two important enzymes, desaturase and elongase, that work to increase the degree of unsaturation. Elongase works by adding carbon atoms into the chain, while desaturase works to introduce double bonds by removing hydrogen atoms [27]. The n-3 and n-6 FA families compete especially at the rate-limiting Δ6 desaturase enzyme that both pathways start with. Usually desaturase enzymes display highest affinity to ALA (n-3 family) more than LA (n-6 family) [28]. The n-3 pathway starts with ALA (C18:3n-3) and ends with DHA (C22:6n-3), while the n-6 pathway starts with LA (C18:2n-6) and ends usually with AA (C20:4n-6) [27, 29].

The major difference between n-3 and n-6 pathways is that n-6 pathway usually does not proceed beyond AA; however, the n-3 pathway involves more complex steps. In the n-3 pathway, there are two elongation steps after the formation of EPA (C20:5n-3), leading to the formation of tetracosapentaenoic acid (C24:5n-3), followed by the reduction by Δ6 desaturase enzyme to produce tetracosahexaenoic acid (Nisinic acid) (C24:6n-3). The final step in n-3 pathway yields DHA (22:6n-3) by the retroconversion of (24:6n-3) to (22:6n-3) which involves peroxisomal ß-oxidation step (**Figure 2**) [27, 30]. In the mammalian cells, the FAs from (n-3) and (n-6) families cannot be interconverted because of lack of Δ12 or Δ15 desaturase enzymes, while the interconversion step takes place in plants [28].

1.5 Sources of omega-3 PUFAs

Fish is a major source of animal protein diet in many countries that is easily digestible than red meat because fish flesh contains long muscle fibers [18]. Consumption of fish have many benefits for human health due to its high content of essential n-3 PUFAs, namely, EPA and DHA [14, 31, 32].The FAs content of fish varies according to diet, i.e., availability of planktons [33], environmental conditions

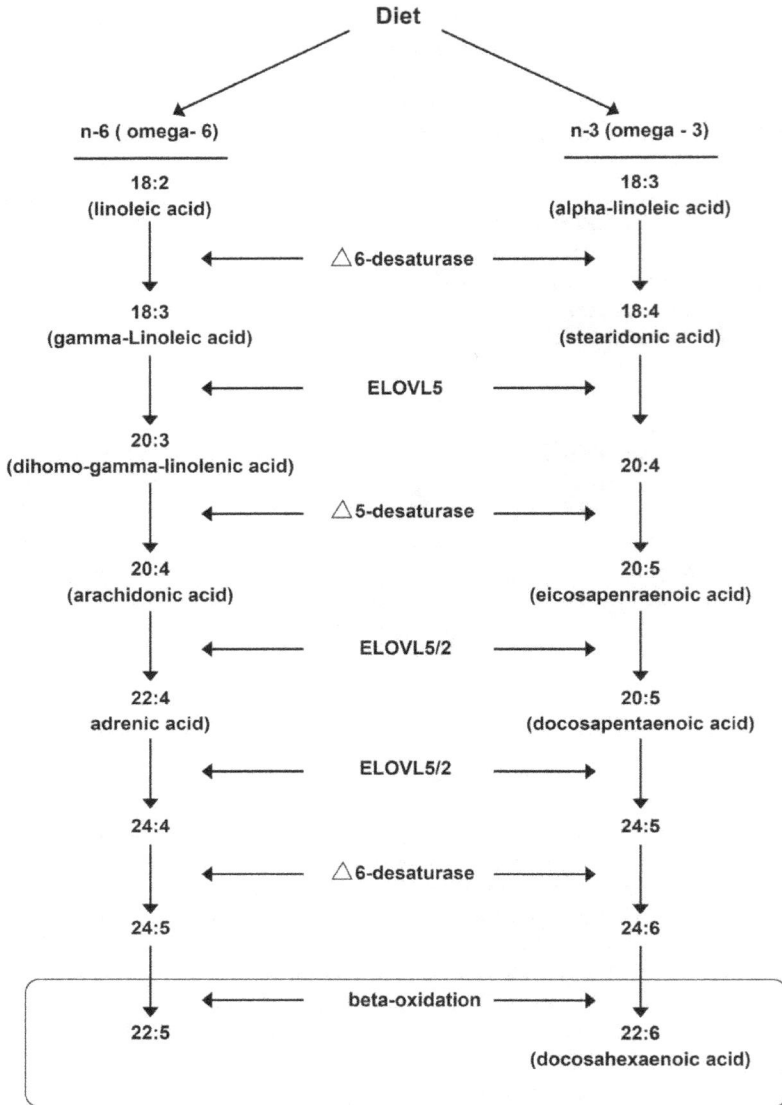

Figure 2.
The elongation-desaturation pathway for the metabolism of n-6 and n-3 polyunsaturated fatty acids [30].

[15], and seasonal variation [34]. A regular consumption of fatty fish prevents cardiovascular disease and neural disorders [35].

PUFAs are important for brain and retina development since studies with animals have shown that deficiencies in n-3 FAs decrease the concentration of DHA in the brain and retina tissues [16]. Fish and terrestrial animals are not able to synthesize n-3 or n-6 PUFAs. The primary producers for PUFAs are marine photosynthetic organisms including phytoplankton, macroalgae, and seaweeds [36].

Only some microalgae species are effective producers for EPA and DHA; therefore, aquatic ecosystems are the principal sources of these two essential PUFAs in the biosphere, in which humans obtain these FAs through the consumption of fish and other marine products (**Figure 2**) [30, 37]. Fish need PUFAs to tolerate low water temperature; thus, low amounts are expected in wormer water like tropical waters [38]. Fish oil become very popular and plays an important role in the prevention of cardiovascular disease and some type of cancers like breast, colon, and prostate [39]. It was observed that there are lower incidence of cardiovascular diseases, hypertension, and autoimmune disorders in populations that consume diet rich in marine fish like Eskimos and Japanese [40].

2. Fish fermentation in relation to fatty acids

Fresh fish is prone to spoilage which is caused by both microbiological and chemical reactions [41]. Lipid deterioration limits the shelf life of oily fish during storage [42]. Lipid hydrolysis induced by lipases and phospholipases produce free FAs that are susceptible to further oxidation to produce low molecular weight compounds responsible for the rancid of fish products [43]. Fish quality may decline during processing and storage mainly due to the oxidation of the PUFAs which is related to the production of unpleasant flavors and odors [44, 45]. EPA and DHA are especially susceptible to oxidation during heating or other culinary treatments [46].

A long time ago, people used to preserve food either by sun drying or salting methods [47]. Salting of fish is an old-age technology that is still in use nowadays even in developing countries due to its simplicity and low cost [47, 48]. Salting fatty fish involves a certain degree of fermentation which is brought by autolytic enzymes from the fish and microorganisms in the presence of high concentrations of salt [47]. Fish fermentation is the transformation of organic substances into simpler compounds like peptides and amino acids by the action of endogenous enzymes or microorganisms, normally in the presence of salt [48–50].

Lactic acid bacteria (LAB) are used to ferment fish along with other food materials like dairy, meat, vegetables, and beverage products [51] resulting in shelf life extension and the addition of new aromas and consistencies [32]. Fermentation by lactic acid bacteria preserves food by the production of lactic acid and other organic acids, which help to reduce pH of the food and inhibit the growth of pathogenic and spoilage organisms [52].

2.1 Fish fermentation all over the world

2.1.1 Asian fermented fish products

Many types of fish sauce and paste are famous in Japan, Southeast Asia, and India. Traditional fermented Japanese seafood is Hatahata-zushi which is processed with boiled rice. Hatahata-zushi is prepared by soaking the sandfish (*Arctoscopus japonicas*) in water, salt, and rice vinegar, which is then fermented with boiled rice and some vegetables [53]. Bagoong is a fish paste produced in the Philippines through the neutral fermentation process of whole fish or shrimp in the presence of 20–25% salt, while Patis is a Philippine yellowish fish sauce prepared from sardines, anchovies, and shrimps [33, 50]. "Suanyu" is a Chinese low-salt fermented whole fish snack which is prepared from fresh fish mixed with cooked carbohydrates, salt, and spices [54]. Lona ilish is a salt fermented fish from India which has strong aroma mixed with some sweet, fruity, and acidic flavors with some saltiness [47].

2.1.2 European fermented fish products

Scandinavia is the main producer of fermented fish products in Europe. Surstromming is produced in Sweden and rakefish in Norway. These fermented fish are made by immersing whole herring and trout in brine (salty water) for 1–2 days, eviscerating, and retaining the roe or milt in barrels with fresh brine. The final fermented product is packed in cans and usually consumed on special occasions [49]. "Tidbits" is another Scandinavian product that is canned with vinegar, sugar, and spices after maturation. In France, the fish *Engraulis encrasicholus*is is salted to prepare anchovies, while in southern France, a fish sauce called "Pissala" is prepared from small fish of the *Engraulis* sp., *Aphya* sp., and *Gobius* sp. [49].

2.1.3 Middle East, African, and Asian fermented fish products

Many types of fermented fish are prepared in the Middle East. "Fseekh" is a famous fermented fish prepared in Egypt and Sudan from different types of fish [49]. Kejeik is a fermented dried fish produce that is common in Sudan and central Africa which is powdered and thickened with okra after boiling [48]. The FA content of Bouri fish varies with fish size; in which it was much greater in small size than large size fish because of the higher activity of lipolytic enzymes in small fish as stated by the authors [48]. However, the fresh and fermented product Fseekh of both fish sizes had the same FA profile. The high salt content was found to have no effect on those enzymes responsible for the liberation of free fatty acids from the lipids [48]. The ratio of UFAs/SFAs decreased as the amount of UFAs decreased significantly after the salting and fermenting process. Moreover, all SFAs except palmitic acid (C16:0) increased significantly. The major SFA was C16:0, and the major UFAs were palmitoleic acid (C16:1n-7) and oleic acid (C18:1n-9). Appreciable amounts of PUFAs such as C18:2n-6, C18:3n-3, stearidonic acid (C18:4n-3, SDA), C20:5n-3, docosapentaenoic acid (C22:5n-3, DPA), and C22:6n-3 were also present. The presence of the odd-chain pentadecylic acid (C15:0) and heptadecanoic acid (C17:0) FAs is considered a unique characteristic of the Bouri fish oil [48].

The analysis of fatty acid compositions was done for different fermented seafoods all over the world. For example, the analysis of Hatahata-zushi which is a Japanese fermented fish product of sandfish has shown no change in the fatty acid compositions throughout the fermentation process; however, the content of SFAs and MUFAs increased, while the content of PUFAs decreased markedly during the process of fermentation. The highest concentrations of the FAs in "Hatahata-zushi" were C18:1n-9, C16:0, C22:6n-3, C20:5n-3, and C16:1n-7, respectively [53].

It has been noticed that bacterial enzymes play an important role in fish fermentation in which the aroma in fermented fish products is claimed to be derived from the activity of halophilic bacteria [47]. The enzymes that participate in the production of PUFAs are thought to be either fish tissue enzymes or microbial enzymes [47]. The traditional fermented fish product Lona ilish of Northeast India has shown that salting plays an important role in the fermentation and preservation of food. Salt preserves the fermented products through the reduction of water activity (aw) of the system, thus rendering a condition less suitable (low moisture) for the microbial growth. Generally, food pathogenic bacteria are inhibited when the water activity is 0.92 or less which is equivalent to NaCl concentration of 13% (w/v). It was reported that halophilic bacteria were responsible for the fermentation process of fish product Lona ilish since these bacteria can tolerate high salt conditions. The bacterial flora of *Micrococcus* and *Bacillus* species were reported to be involved in the processing of'Lona ilish [47]. Other microorganisms are found in the spontaneously fermented Chinese fish product Suanyu such as lactic acid bacteria, *Staphylococcus*, and yeast [54].

The two main products of fish fermentation in the Arabian Gulf region are fish paste and sauce. It is difficult to classify these products due to the lack of standardization throughout the world. Generally, fish paste is a thick product with whole fermented fish, while fish sauce is a thinner product with additives such as spices and cereals. The color of fish paste or sauce is usually brown resembling soy sauce. Fish paste is more nutritious than fish sauce [49]. Tareeh is a salt fermented fish paste in the Arabian Gulf countries prepared from a high-fat fish *Sardinella* spp. [33, 49]. The famous fish sauce in the Arabian Gulf Countries is called Mehiawah that is prepared by the addition of spices to the fermented fish Tareeh [49, 55].

The fatty acid contents of the fish product Tareeh and Mehiawah of white sardinella (Oom) were recently investigated [56, 57]. The study of Freije et al. (2018) was conducted in order to determine the effect of fermentation under high salting conditions (20%) for 8 weeks on weekly basis on the fatty acid compositions of white sardinella (Sardinella albella) [57]. The fatty acid compositions of Tareeh have shown a great deal of variation as the fermentation process proceeded. The concentrations of SFAs and MUFAs significantly declined in Tareeh samples compared to raw fish, whereas the concentrations of UFAs and PUFAs were significantly increased starting from week 4 of fermentation. The amount of n-6 FAs as well as n-3 FAs increased during the fermentation process, and the enzymatic activities of the elongases and desaturases were found to have higher affinity to n-3 FAs than n-6 FAs. The ratio of n-6/n-3 was decreased after 8 weeks of fermentation, while the proportions of EPA + DHA were increased. This unique fermentation process might have a great application in the food industry [56]. It was also concluded that the elongation and desaturation process was carried out by single new bacterial strain from the *Bacillus* species named *Bacillus mojavensis-ASK* [58, 59].

3. Conclusion

The studies that investigated fatty acid compositions in fermented fish products all over the world are limited. Each of those studies has given different results without any common consistency among them. Some of those studies have reported a decrease in UFAs and an increase in SFAs, while others reported increased SFAs and MUFAs but decreased PUFAs during fermentations. On the contrary, the most recent studies have reported declined levels of SFAs and MUFAs and a rapid increase in PUFAs. Such contradicted results can be attributed to the difference in the type of fermented fish, fermentation conditions, and the addition of different additives. Therefore, this process requires thorough investigations in order to reveal the mystery behind such contradictions.

Acknowledgements

The authors are grateful to the Department of Biology, College of Science, University of Bahrain, where the study was conducted and financially supported.

Author details

Afnan Freije[1*] and Aysha Mohamed Alkaabi[2]

1 Department of Biology, College of Science, University of Bahrain, Kingdom of Bahrain

2 University of Bahrain, Sakhir, Kingdom of Bahrain

*Address all correspondence to: afreije@uob.edu.bh

IntechOpen

References

[1] Karp G. Cell and Molecular Biology. 5th ed. Vol. 42. Hoboken: John Wiley & Sons, Inc; 2008. pp. 47-49

[2] McMurry J, Simanek E. Fundamentals of Organic Chemistry. 6th ed. Thomson Brooks/Cole: Belmont; 2007. pp. 510-513

[3] Mader SS. Biology. 12th ed. New York: McGraw-Hill companies; 2013. pp. 42-46

[4] Fahy E, Subramaniam S, Brown HA, Glass CK, Merrill AH Jr, Murphy RC, et al. A comprehensive classification system for lipids. The Journal of Lipid Research. 2005;**46**:839-862

[5] Wakil SJ. Mechanism of fatty acid synthesis. Journal of Lipid Research. 1961;**2**(1):1-24

[6] Leonard AE, Pereira SL, Sprecher H, Huang YS. Elongation of long-chain fatty acids. Progress in Lipid Research. 2004;**43**(1):36-54

[7] Boyer R. Lipids, Biological Membranes, and Cellular Transport In Concepts in Biochemistry. 1st ed. Pacific Grove: Brooks/Cole Publishing Company A division of International Thomson Publishing Inc.; 1999. pp. 238-275

[8] Boyer R. Lipids, biological membranes, and cellular transport. In: Concepts in Biochemistry. 2nd ed. Canada: John Wiley & Sons, Inc.; 2002. pp. 208-240

[9] Campbell MK, Farrell SO. Lipids and proteins are associated in biological membranes. In: White A, Olafsson J, Bowen L, Weber L, Van Camp S, editors. Biochemistry. 7th ed. China: Brooks/Cole, Cengage Learning; 2012. pp. 193-225

[10] Hames D, Hooper N. Lipid Metabolism in Biochemistry. 3rd ed. Cornwall: Taylor & Francis Group; 2005. pp. 335-371

[11] McGuire M, Beerman KA. Lipids in Table of Food Comparison for Nutritional Sciences from Fundamentals to Food. 3rd ed. Belmont: Wadsworth, Cengage Learning; 2013. pp. 217-268

[12] Sizer F, Whitney E. The Lipids: Fats, Oils, Phospholipids, and Sterols In Nutrition Concepts and Controversies. 12th ed. USA: Wadsworth, Cengage Learning; 2011. pp. 150-188

[13] Shireman R. Essential Fatty Acids in Encyclopedia of Food Sciences and Nutrition, 2nd ed. London: Academic Press; 2003. pp. 2169-2176

[14] Gladyshev MI, Sushchik NN, Anishchenko OV, Gubanenko GA, Demirchieva SM, Kalachova GS. Effect of boiling and frying on the content of essential polyunsaturated fatty acids in muscle tissues of four fish species. Food Chemistry. 2007;**101**:1694-1700

[15] Mehta LR, Dworkin RH, Schwid SR. Polyunsaturated fatty acids and their potential therapeutic role in multiple sclerosis. Nature Clinical Practice. Neurology. 2008;**5**:82-92

[16] Rebah FB, Abdelmouleh A, Kammou W, Yezza A. Seasonal variation of lipid content and fatty acid composition of *Sardinella aurita* from the Tunisian coast. Journal of the Marine Biological Association of the United Kingdom. 2010;**90**(3):569-573

[17] Bulla MK, Simionato JI, Matsushita M, Coró FAG, Shimokomaki M, Visentainer JV, et al. Proximate composition and fatty acid profile of raw and roasted salt-dried sardines (*Sardinella Brasiliensis*). Food and Nutrition Sciences. 2011;**2**:440-443

[18] Valenzuela R, Videla LA. The importance of the long-chain polyunsaturated fatty acid n-6/n-3 ratio in development of non-alcoholic fatty

liver associated with obseity. Food & Function. 2011;**2**:644-648

[19] Louly AWOA, Gaydou EM, ElKebir MVO. Muscle lipids and fatty acids profiles of three edible fish from the Mauritanian coast: *Epinephelus aeneus*, *Cephalopholis taeniops* and *Serranus scriba*. Food Chemistry. 2011;**124**:24-28

[20] Njinkoué JM, Barnathan G, Miralles J, Gaydou EM, Samb A. Lipids and fatty acids in muscle, liver and skin of three edible fish from the Senegalese coast: *Sardinella maderensis*, *Sardinella aurita*, and *Cephalopholis taeniops*. Comparative Biochemisty and Physiliogy Part B. 2002;**131**:395-402

[21] Hirafuji M, Machida T, Hamaue N, Minami M. Cardiovascular protective effects of n-3 polyunsaturated fatty acids with special emphasis on docosahexaenoic acid. Journal of Pharmacological Sciences. 2003;**92**(4):308-316

[22] Dyerberg J, Bang HO. Haemostatic function and platelet polyunsaturated fatty acids in Eskimos. Lancet. 1979;**2**(8140):433-435

[23] Ward OP. Microbial production of long-chain polyunsaturated fatty acids. Inform. 1995;**6**:683-688

[24] Echeverria F, Valenzuela R, Hernandzez-Rodas MA. Docosahexaenoic acid (DHA), a fundamental fatty acid for the brain: Newdietry sources. Prostaglandins, Leukotrienes, and Essential Fatty Acids. 2017;**124**:1-10

[25] Barrera C, Valenzuela R, Chamorro R, Bascunan K, Sandoval J, Sabag N, et al. The impact of maternal diet during pregnancy and lactation on the fatty acid composition of erythrocytes and breast milk of Chilean women. Nutrients. 2018;**10**:839. DOI: 10.3390/nu10070839

[26] ISSFAL: Issfal Board Statement No.3. 2004. Available from: <www.issfal.org.uk>

[27] Visioli F, Richard D, Bausero P, Galli C. Role of Polyunsaturated Omega-3 Fatty Acids and Micronutrients Intake on Atherosclerosis and Cardiovascular Disease in Nutrition and Metabolic Bases of Cardiovascular Diseases. West Sussex, UK: Blackwell Publishing Ltd.; 2011. pp. 166-175

[28] Johns PJH, Kubow S. Lipids, Sterols, and their Metabolism in Modern Nutrition in Health and Diseases. 10th ed. Lippincott Williams & Wilkins; 2006. pp. 92-122

[29] Thomas BJ. Efficiency of conversion of α-linolenic acid to long chain n-3 fatty acids in man. Lipid metabolism and therapy. 2002;**5**(2):127-132

[30] Edwards IJ, O'Flaherty JT. Omega-3 fatty acids and PPARγ in cancer, PPAR Research. 2008. Available from: http://www.hindawi.com/journals/ppar/2008/358052/ (Accessed: August, 2019)

[31] Gladyshev MI, Sushchik NN, Anishchenko OV, Makhutova ON, Kalachova GS, Gribovskaya IV. Benefit-risk of food fish intake as the source of essential fatty acids vs. heavy metals: A case study of Siberian grayling from the Yenisei River. Food Chemistry. 2009;**115**:545-550

[32] Nordvi B, Egelandsdal B, Langsrud Ø, Ofstad R, Slinde E. mDevelopment of a novel, fermented and dried saithe and salmon product. Innovative Food Science & Emerging Technologies. 2007;**8**:163-171

[33] Al-Jedah JH, Ali MZ, Robinson RK. The nutritional importance of local communities of fish caught off the coast of Qatar. Nutrition and Food Science. 1999;**99**(6):288-294

[34] Egan H, Kirk SR, Sawyer R. Pearson's Chemical Analysis of Food, 8th ed. London: Chrchill Livingstone; 1981

[35] Slivers KM, Scott KM. Fish consumption and self-reported physical and mental health status. Public Health Nutrition. 2002;**5**:427-431

[36] Servel MO, Claire C, Derrien A, Coiffard L, De-RoeckHoltzhauer Y. Fatty acid composition of some marine microalgae. Phytochemistry. 1994;**36**:691-693

[37] Arts MT, Ackman RG, Holub BJ. Essential fatty acids in aquatic ecosystems: A crucial link between diet and human health and evolution. Canadian Journal of Fisheries and Aquatic Sciences. 2001;**58**:122-137

[38] Bolgova DM, Bogdan VV, Ripatti PO. Effects of the temperature factor on fish fatty acid composition. In: Sidorov VS, Vysotskaya RU, editors. Sravn. Biokhim. Vodn. Zhivotn. Comparative Biochemistry of Aquatic Animals, Petrozavodsk: Karel. Filial AN SSSR. 1983. pp. 52-61

[39] Mori TA. Omega-3 fatty acids and hypertension in humans. Clinical and Experimental Pharmacology & Physiology. 2006;**33**:842-846

[40] Lands WEM. Fish and Human Health. London: Academic Press; 1986

[41] Chaijan M, Benjakul S, Visessanguan W, Faustman C. Changes in lipids in sardine (*Sardinella gibbosa*) muscle during iced storage. Food Chemistry. 2006;**99**:83-91

[42] Cho S, Endo Y, Fujimoto K, Kaneda T. Oxidative deterioration of lipids in salted and dried sardines during storage at 5°C. Nippon Suisan Gakkaishi. 1989;**55**:541-544

[43] Toyomizu M, Hanaoka K, Yamaguchi K. Effect of release of free fatty acids by enzymatic hydrolysis of phospholipids on lipid oxidation during storage of fish muscle at −5°C. Bulletin of the Japanese Society of Scientific Fisheries. 1981;**47**:605-610

[44] Ackman R. Fatty acids. In: Ackman RG, editor. Marine Biogenic Lipids, Fats and Oils. Vol. 1. Boca Raton, FL: CRC Press; 1989. pp. 103-137

[45] Hsieh R, Kinsella J. Oxidation of polyunsaturated fatty acids: Mechanisms, products, and inhibition with emphasis on fish. Advances in Food Research and Nutritional Research. 1989;**33**:233-341

[46] Candela M, Astiasaran I, Bello J. Deep-fat frying modifies high-fat fish lipid fraction. Journal of Agricultural and Food Chemistry. 1998;**46**:2793-2796

[47] Majumdar RK, Basu S. Characterization of the traditional fermented fish product Lona ilish of Northeast India. Indian Journal of Traditional Knowledge. 2010;**9**(3):453-458

[48] El-Sebaiy LA, Metwalli SM. Changes in some chemical characteristics and lipid composition of salted fermented Bouri fish muscle (*Mugil cephalus*). Food Chemistry. 1988;**31**:41-50

[49] Al-Jedah JH, Ali MZ. Fermented fish products in Encyclopedia. Food Microbiology. 1999:753-759

[50] Peralta EM, Hatate H, Kawabe D, Kuwahara R, Wakamatsu S, Yuki T, et al. Improving antioxidant activity and nutritional components of Philippine salt-fermented shrimp paste through prolonged fermentation. Food Chemistry. 2008;**111**:72-77

[51] Campbell-Platt G. In Fermented Foods of the World: A Dictionary and Guide. Kent, England: Butterworths; 1987. pp. 157-158, 173-174, 202

[52] Vandenberfh A. Lactic acid bacteria, their metabolic products and interference with microbial growth. FEMS Microbiological Reviews. 1993;**12**:221-238

[53] Chang CM, Ohshima T, Koizumi K. Changes in the composition of free amino acids, organic acids and lipids during processing and ripening of 'Hatahata-Zushi', a fermented fish product of sandfish (*Arctoscopus japonicas*) and boiled rice. Journal of the Science of Food and Agriculture. 1994;**66**:75-82

[54] Zeng X, Xia W, Jiang Q, Yang F. Chemical and microbial properties of Chinese traditional low-salt fermented whole fish product Suanyu. Food Control. 2013;**30**:590-595

[55] Al-Laith AA. Fate of *Staphylococcus aureus* in experimentally prepared fermented fish sauce (Mehiawah). Arab Gulf Journal of Scientific Research. 2008;**26**(4):184-192

[56] Freije A. Fatty acid compositions of fish white Sardinella (Oom), fish paste (Tareeh), and fish sauce (Mehiawh): Fermented fish products rich in polyunsaturated fatty acids. Bahrain Medical Bulletin. 2017;**39**(4):200-209

[57] Freije A, AlKaabi A, Al-Thawadi S, Saleh K, Bin Thani A. Changes in the fatty acid compositions of Tareeh fermented fish product of *Sardinella albella* locally known as Oom in the Kingdom of Bahrain. Ecology, Environment, and Conservation. 2018;**24**(3):19-28

[58] Al-Thawadi S, Saleh K, Freije A, Bin Thani A, Al-Kaabi A. Isolation, identification and characterization of a *Bacillus* species isolated from a common fermented fish product (Tareeh) of white sardinella (*Sardinella albella*) in the middle east in relation to fatty acids profile: Elongase and desaturase activities. Asian Journal of Microbiology, Biotechnology & Environmental Sciences. 2017;**19**(3):1-12

[59] Bin Thani A, Al-Thawadi S, Freije A. Structure-function analysis of the multi-domain beta-ketoacyl synthase involved in the production of Eicosapentaenoic acid, docosahexaenoic acid and antibiotics by in-silico comparative approaches and phylogenetic analysis. Asian Journal of Microbiology, Biotechnology & Environmental Sciences. 2017;**19**(3):1-9

www.ingramcontent.com/pod-product-compliance
Lightning Source LLC
Chambersburg PA
CBHW081233190326
41458CB00016B/5759